Have you any wool?

Have you any wool?
THE CREATIVE USE OF YARN

JAN MESSENT

SEARCH PRESS

First published in Great Britain 1986
by SEARCH PRESS LTD
Wellwood, North Farm Road
Tunbridge Wells, Kent
TN2 3DR

ISBN 0 85532 566 6

Note on the illustrations

For clarity and ease of reference, all the illustrations have
been numbered consecutively through the book and
described as 'Figures' whether they are diagrams, figures
or photographs.

Origination by Adroit Photo Litho, Birmingham.
Printed in Spain by Artes Graphicas Elkar, S. Coop.
Autonomía, 71 - 48012-Bilbao - Spain

Contents

Introduction

This is a knitting/crochet book with a difference, for it deals with the learning-process in an imaginative way, intended to make what may be a completely new skill into sheer delight instead of sheer misery. And how many of us can remember the frustrations of having to 'do it right'? There is very little emphasis, in this book, on 'doing it right', though the correct ways are there for all to look at if they wish, or to ignore if they prefer. The basic stitches of knitting and crochet are explained, fully and simply, side by side, with no emphasis on the superiority of either, but a hope that they will be seen just as alternative ways of making a fabric, with their own unique advantages and disadvantages. It is to this end that I have purposely avoided giving too many cut-and-dried instructions, hoping that the reader will mix knitting and crochet together with equal ease, as I do, depending on the effect required.

For far too long, knitters and crocheters have relied on exact instructions for every project, no matter how small and simple, never daring to stray into the realms of fancy yarns in case it gets 'too complicated'. This book aims to encourage people to use small quantities of any and every yarn available to them for small interesting and decorative projects, for it is only by *using* the yarns in this way that we shall actually discover what they – and we – can do. Not only this, but we shall have spent very little time, money and effort in the process, and have gained experience, great enjoyment and a piece of craftwork to provoke delight and admiration.

The beauty of these projects is that they can be made simple or complicated according to the ability and imagination of each reader, which is why the book is not *specifically* aimed at children but at anyone who enjoys knitting and crochet for their own sake, not merely as something to be got through in order to have a new garment. Garments are not discussed here; there are hundreds of books and publications which cover that aspect of the craft more than adequately. To involve complete beginners in lengthy and costly garment-making activities seems, to me, rather like being asked to write an essay the day after you have been introduced to the alphabet. A certain amount of play time, experimental time, call it what you will, is required first, to become acquainted with the various tools, yarns and stitches, a certain time to acquire that familiarity and confidence with the techniques which contributes towards a pleasant and rewarding activity, which removes that mystery so long attributed to knitting and crochet by those who do not!

In describing some of the stitches (e.g. knitted bobbles), I have attempted to explain the principal idea behind the working method rather than write line-by-line instructions, in the hope that this will lead to a better understanding of what is happening. By doing this, I would like to think that one would be stimulated into an exploration of yarns, stitches and tools to discover new ideas. Plainly, this book is not for the faint-hearted: most of the projects here demand a few basic decisions. Which needles? Which yarns? Which stitch? On the other hand, I have always tried to give enough information about the projects to enable the reader to make a sensible choice, so that no-one will find that they have made an owl as big as a chimney-pot instead of a salt-pot! If you read the chapter about facts and fibres, and about tools, then you will know what to do, and as your project will not have two sleeves of the same length, neither will it need to go over your head and fit properly – it really won't matter at all whether it is slightly bigger, smaller, thinner or thicker than you expected it to be. For young people, this is an essential learning experience, and fun too.

Readers may notice that my use of certain words, and the general tone of the text, is also geared towards various age-ranges, on the assumption that the very simplest experiments will be more favoured by very young people, and the more sophisticated ones by older readers. Hence my use of both 'oblong' and 'rectangle', for instance. I am sure that many older readers will enjoy the small projects every bit as much as the younger ones, therefore in my resort to very simply-written instructions I am merely trying to make it easy for the latter category, not to patronise the former.

Nearly all the work shown here now belongs to the Knitting Craft Group, an organisation set up and sponsored by the members of the British Hand Knitting Association. It is reproduced here by the courtesy of its Director, Alec Dalglish, who has been more than co-operative in so many ways throughout my work on this book. It is to him that I owe special thanks not only for his help, but for opening the door into the 'Aladdin's Cave of yarn' which has transported my teaching and embroidering experiences into another world. Now, my involvement with knitting and crochet looks like keeping me entangled until I can get the genie back into his lamp. And you can imagine how difficult *that* is!

The photographs are nearly all the work of Richard Hill of Jigsaw Studios, Batley, West Yorkshire, to whom much credit is due. The last piece to be made for the book was the Butterfly on page 93: he is therefore not familiar with the other material and has made repeated attempts to escape among the pages to find out more about it, behaving with complete irreverence, and thus setting the tone for all of us.

CHAPTER ONE

Facts and fibres

Fibres, ply, thickness and texture – The ball band –
Pompons, tassels and finger-cords – Yarn-colour and dyes –
Small-scale weaving projects – French knitting – Needle-
weaving – Card-wrapping with yarn

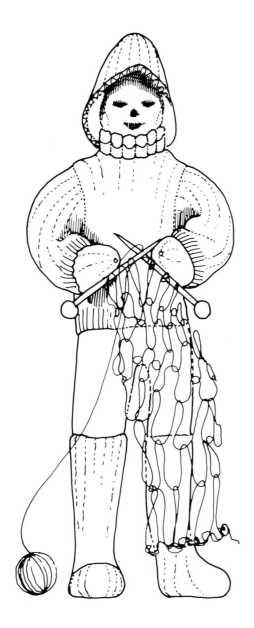

*Alec the fisherman is knitting a net to catch all would-be
knitters and crocheters so that he can infect them with his
enthusiasm.*

Fig. 2.

Facts and fibres

FIBRES

If you are new to the idea of creating with yarn, you will almost certainly need to know what the various types of wool are made of and why they behave differently from each other. After all, this can make a lot of difference to a) the cost, b) the final result of your project, c) your temper, and d) your willingness (or otherwise) to use yarns again.

The three main sources of fibres from which yarns are made are animals, plants and synthetics, the last meaning that they are produced from chemicals in laboratories, sometimes including metal for glitter effects. Many of the yarns we use may be made of different fibres blended together; some yarns may be made of only one of them, for example, wool or acrylic, sometimes two, like acrylic and nylon, and sometimes as many as three or four may be used together. Each fibre contributes its own quality to create softness, strength, fluffiness and sheen.

All these different fibres and their blends are produced in different thicknesses, and in different textures too. The sources of the various fibres are shown here, and are listed below (see Fig. 2):

1. Sheep: produces wool for all purposes.
2. Silk-moth: the cocoon produces silk, a fine, shiny thread, often mixed with other fibres.
3. Angora rabbit: produces long, silky and soft hair.
4. Cotton: a fibre from the soft, downy tops of the cotton bush, used alone, or blended with other fibres.
5. Flax: produces linen from the stems of the plant.
6. Wood: used in the manufacture of viscose rayon, acetate and tri-acetate. A man-made fibre.
7. Man-made fibres from synthetic polymers, e.g. acrylic, nylon and polyester.
8. Camel: produces a soft, warm hair usually sold in the natural colours.
9. Llama: a relation of the camel, produces fine hair, usually sold undyed.
10. Alpaca: another of the camel family produces a fine, expensive hair, usually mixed with other fibres.
11. Vicuna: a smaller relative from Peru; wild, produces extremely soft, warm and light hair but difficult to obtain, and very expensive.
12. Angora goat: produces mohair; very light and warm, often sold brushed.
13. Himalayan goat: produces cashmere, wonderfully light, warm and soft but very expensive.

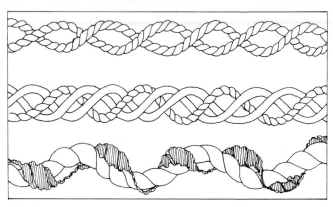

Fig. 3.

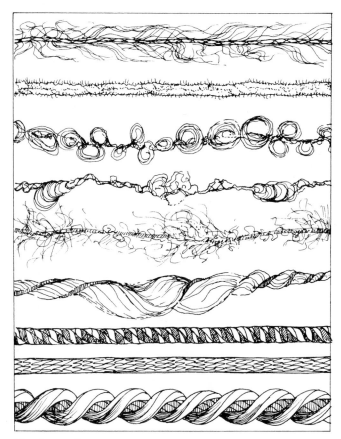

Fig. 4.

Ply and thickness

Not all yarns are spun; some are woven into fine tubes, and some very thick yarns are barely twisted at all, although strands of twisted yarns are usually twisted together or 'plied' to make one strand. The number of strands used together is called the 'ply' – it may be one, two, three or four ply. However, just to make things more confusing, this will not always give you an indication of how thick a yarn is, as a two-ply *may* be thicker than a four-ply. It all depends on how thick are the individual strands.

Generally speaking, though, we usually refer to the thicknesses of yarns in the following terms as these are generally accepted as the nearest one can get to describing the thickness of a yarn.

Two-ply: very fine and soft.
Three-ply: quite fine.
Four-ply: only a little thicker than three-ply.
Double-knit (DK): by far the largest range of yarns comes in this thickness: quite thick, but not all double-knit is exactly the same.
Aran: a smooth yarn, a little thicker than DK, mostly in creamy-whites.
Double-double Knit: thicker than DK.
Chunky-knit: very thick, sometimes furry.

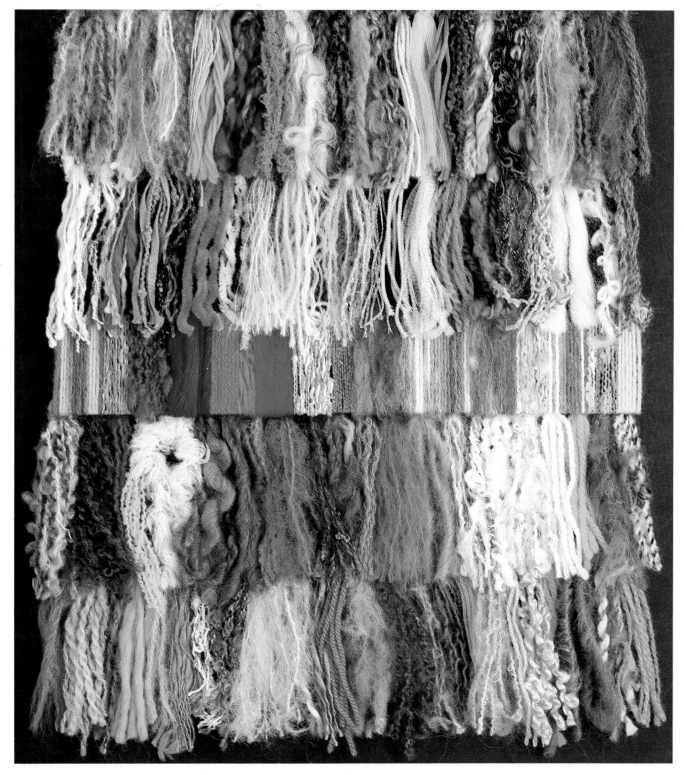

Fig. 5. A selection of yarns to be found in wool shops.

9

Texture

With a combination of different fibres, plies and thicknesses, the spinners can also do many other things to make the texture of yarns more interesting. They can introduce bumps (known as 'slubs'), coils and loops, twists and gimps, metallic chips and flecks of other fibres to make many exciting effects. A few of these types are listed below, which you may be able to identify in the photograph on the previous page:

Woollen spun: soft, slightly hairy.

Brushed: very hairy, often contains mohair, not easy for beginners to use.

Chenille: a velvet-like yarn made of short fibres trapped in a central core.

Bouclé: a looped or bumpy yarn giving a highly-textured surface.

Slub: a textured yarn, with lumps and loops of fibres at irregular or regular intervals.

Crepe: a smooth yarn with a slightly harder twist, giving a crisp look to the fabric.

The ball band

This is the paper band which is wrapped around each ball of yarn; it may give you as many as thirteen pieces of information:

1. The name of the yarn, usually the largest lettering.
2. The name of the manufacturer.
3. The manufacturer's address, and the country where the yarn was made.
4. The quantity/weight in each ball (e.g. 25g, 50g or 100g).
5. B.S.984.76 which means British Standard 984, as legislated in 1976, and refers to the weight of the yarn at a controlled degree of humidity.
6. What the yarn is made of, with the % of each fibre.
7. The colour number.
8. The dye-lot number.
9. Laundering instructions.
10. Recommended sizes of knitting needles and crochet hooks.
11. How many stitches and rows it will knit to the square inch or square centimetre.
12. If a yarn can be washed in a washing machine, it may say 'Machine washable'.
13. If a yarn is made of 100% wool, it may show the Pure New Wool symbol.

   Look at the ball band of any yarn that you buy. The information will help you to make the right choice for your needs.

Fig. 6. The Ball band.

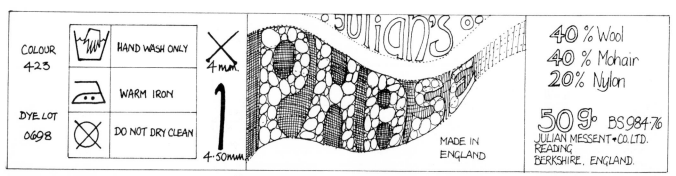

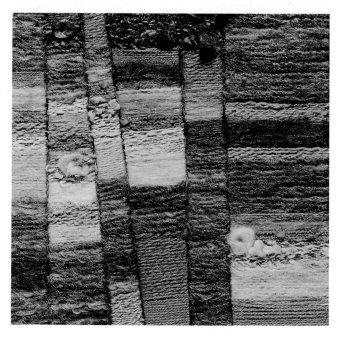

Fig. 7 and Fig. 8. Experiments in yarn texture.

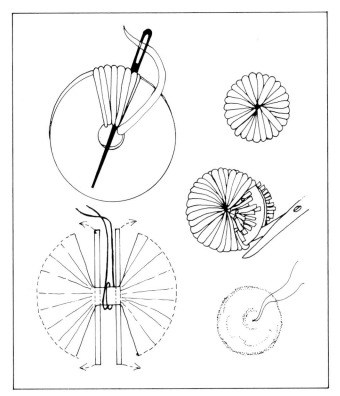

Fig. 9. Pompons.

Card-wrapping experiments

Narrow strips of card have been wrapped with yarns and laid across a larger card on which the threads are wrapped in the other direction. Using different textures in tones of the same colour, an interesting pattern may be built up. Use double-sided sticky-tape along the side of each strip to hold the yarns in place (Fig. 7).

A large piece of card has been cut up into strips of varying widths, some at angles. Each one is wrapped with the same yarns but in a slightly different order, then re-assembled. There are many other variations possible on this theme (Fig. 8).

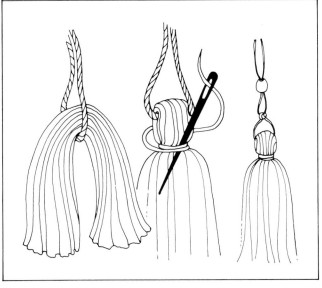

Fig. 10. Tassels.

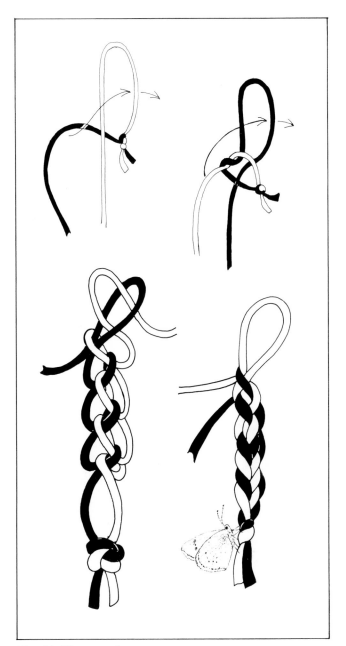

Fig. 11. Finger cord.

POMPONS, TASSELS AND FINGER-CARDS

Pompons (Fig. 9). Cut two discs of card and make a hole in the centre. The size of the disc will determine the size of the finished pompon. Wrap the discs with yarn as shown, until the hole is completely filled, then cut and tie. Multi-coloured pompons are made by using several coloured yarns in layers, or by using random-dyed yarns. Glitter yarns mixed in are particularly effective.

Tassels. Wrap yarns thickly around a stiff piece of card, or a book, then cut and tie as shown in Fig. 10. Use plenty of yarn for this, as thin tassels are not so attractive as thick ones.

Finger-cords (Fig. 11). Knot two differently-coloured thick yarns together and keep hold of the knot in the right hand, and the loop open with the finger and thumb. Pass each cord through this loop alternately, pulling the loop closed each time as shown.

YARN COLOUR AND DYES (Fig. 12)

Yarns may be dyed either before or after spinning, and in many different ways, to produce thousands of beautiful colours and colour-mixtures. Different fibres absorb dyes in different ways, for instance, cotton is much more absorbent than nylon, so yarn made from one fibre will dye in a different way from one made of another. Those yarns which have many blended fibres in them will also react differently, as will yarns with bumps and lumps in, or thick and thin bits.

Words like 'random-dyed', 'heather mixture' and 'spot-print' are used to describe the way some yarns have been coloured. Here is a short list of some of the names you may hear:

1. Solid colour: one colour only, unbroken by any other effects.
2. Flecked: containing tiny flecks of other colours.
3. Metallic: a glitter yarn which may be smooth, or a textured yarn which may contain tiny flecks of metallic bits.
4. Heather mixture: the fibres are dyed in different colours before spinning and then they are mixed and spun together to make a soft blend of colours. Shetland wool is often spun in this way.
5. Marl: single threads of yarn are dyed in different colours and then plied together to give a twisted stripe effect.
6. Print, or Spot-print: the spun yarn is passed through rollers which print two or three different colours on to it in patches. These colours stay quite separate on the yarn and do not blend into each other.
7. Spray-dyed: fine jets of dye are sprayed on to yarn to give soft patches of colour.
8. Random-dyed: different parts of the wet yarn are dipped into different dyes to give quite a long length of colour before it changes to the next one. These colours often merge to make new colours where they meet, and sometimes different tones of the same colour are used in one ball of yarn.

These different ways of dyeing yarn are used on all the various thicknesses, all the yarn-types, all the fibres and the blends of fibres.

As it is not possible to dye all the yarn at once, the colour and dye-lot numbers (which you will find on every ball band) will vary with each batch of yarn packed and sold to the shops. As each new batch is dyed, the colour will vary very slightly and will be given a new dye-lot number, so although the *colour number* may be the same, the dye-lot will be different. You must be quite sure that the yarn you buy for an important project has the same numbers on it, as small changes in colour may show up on the finished article.

MEXICAN, OR GOD'S EYE, WEAVING, TAPESTRY AND FRENCH KNITTING (Fig. 13)

Weaving is probably the oldest known method of producing fabric. The very first pieces were probably quite small, sewn into larger pieces like a patchwork. All the ancient civilizations had their own distinctive type of loom, and used a variety of fibres to produce cloth, including wool, silk, linen and cotton. All types were made in basically the same way, that is, by having vertical threads (the warp) held rigid on some

Fig. 12. Ways of dyeing yarn. The numbered paragraphs in the text correspond to the sections of the yarn.

Fig. 13. Top and right. *Mexican or God's Eye weaving.*
Left. *Needleweaving over a tapestry-woven background.*
Bottom. *French knitting.*

kind of frame, and by weaving another thread (the weft) across these from side to side.

While one type of device, called a loom, will produce a fabric used for clothing, there is another type which is used for making woven pictures, called tapestries. The tapestry shown here is very small, only 6 in.

(15 cm) square, and it was made in two layers, the top one being called 'needleweaving', on a piece of perspex. For needleweaving, see page 15.

Mexican, or God's Eye, weavings are small charms which are said to ward off evil spirits. They hang on the walls of many homes in Mexico, and are sent to friends to bring them good luck.

French knitting is an age-old activity which only needs a wooden cotton-bobbin and four nails, small amounts of yarn and a thick needle. Beautiful and colourful cords can be made very simply by even the youngest child.

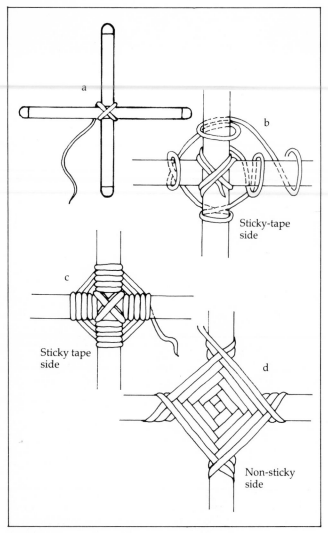

Fig. 14. Mexican, or God's Eye weaving.

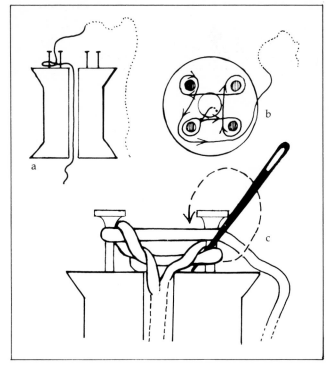

Fig. 15. French knitting.

HOW TO DO MEXICAN, OR GOD'S EYE, WEAVING (Fig. 14)

This is a simple and attractive way of using short lengths of yarn, plain or textured, thick or thin. Both sides of the little hanging are equally decorative; you will be able to decide which one to make the right side as you progress. Multi-coloured, textured and glitter yarns, raffia, string, leather and strips of dyed nylon tights all mix well and produce good results.

Materials

For the smallest God's Eye, you will need two flat lollipop sticks or fine sticks of wood about 4 in. (10 cm) long, and for the larger one you will need two pieces of fine wooden dowelling about 7 in. (18 cm) long. Also, you will need coloured yarns and threads, scissors, a blunt-ended needle and double-sided sticky tape.

Method

Glue two sticks together at right angles, and put two strips of double-sided sticky tape on one side of the sticks, stopping short of the ends (Fig. 14a). Use the first piece of yarn to criss-cross the join in the centre, and begin wrapping as shown in Fig. 14b. Lay the threads neatly and close together, so that they overlap each other slightly. Change yarns by tying knots, or by sticking the ends of the yarns to one arm and covering them with the next threads, or by leaving the long ends to be darned in later on the wrong side. Figs. 14c and 14d show both sides of the work in progress. The little motifs may be decorated with pompons, tassels, bells or beads.

SMALL-SCALE WEAVING PROJECTS

The warp threads may be held taut on a wide variety of items and in several different ways. Some of these ways are shown in Fig. 16 (opposite, right), others not shown include forked branches of wood, polystyrene tiles, shoe-box lids, picture and lampshade frames, metal rings, squares of perspex and all kinds of boxes and thick card.

The warp threads are kept in place either by wrapping thread around pins stuck into the card at the top and bottom edges (a), by cutting V-shapes into the edges and wrapping the thread right round the card (b), or by cutting slits into the card (d); while (c) shows a rigid meat-tray with tiny slits in the edges ready to be warped.

FRENCH KNITTING (Fig. 15)

This is a cord-making process using a cotton bobbin with four nails in the top. A favourite activity among children of all ages, it is also useful for producing decorative cords for garments, belts and straps, especially when several lengths are plaited together. The end of the yarn is dropped down the centre of the bobbin (a) and the yarn is then wrapped around the nails in an anti-clockwise direction (b), although it may actually be done either way. With a blunt needle, pick up the loop already on each nail (c) and pull it up and over the top of the 'new' yarn to make a new stitch. The cord produced should be gently pulled down through the centre hole as work progresses.

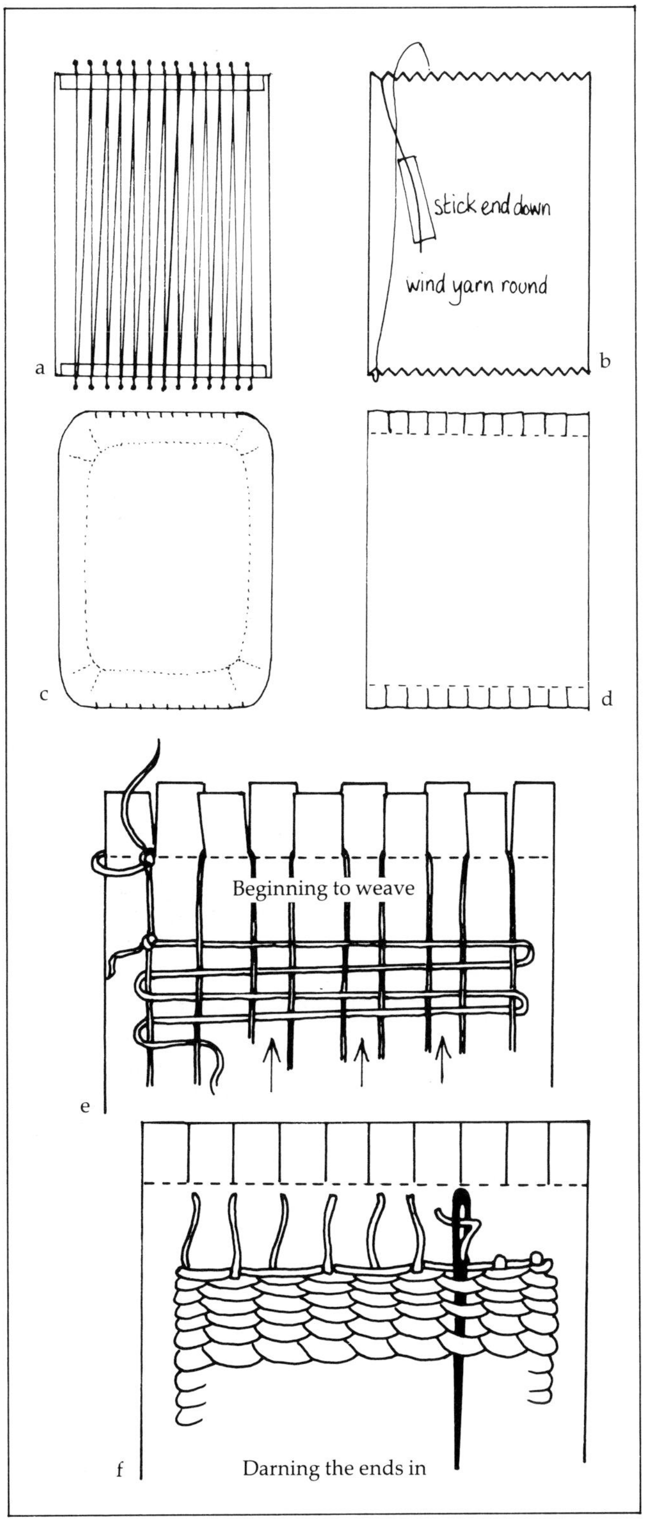
stick end down
wind yarn round
a
b
c
d
Beginning to weave
e
f
Darning the ends in

Fig. 16. Ways of warping for simple weaving projects.

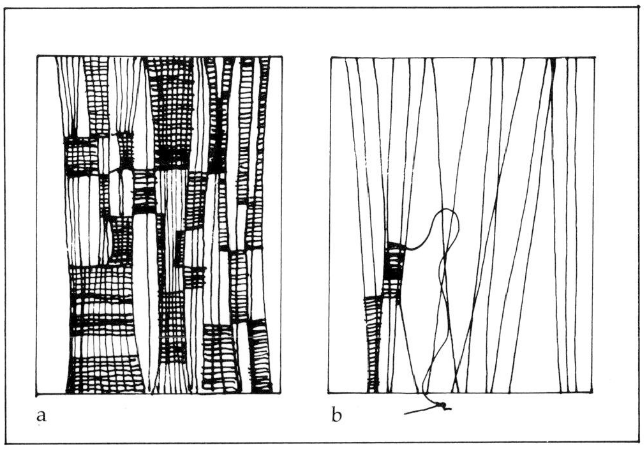
a
b

Fig. 17. Needleweaving on card or perspex.

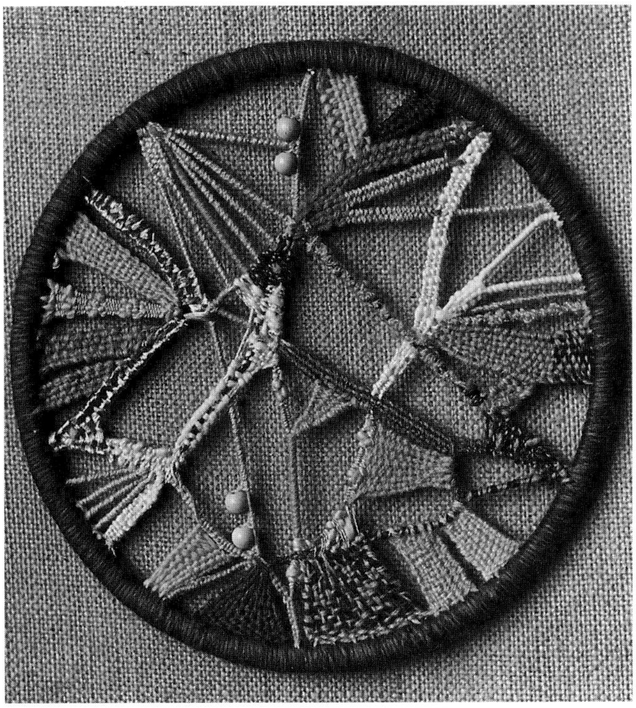

Fig. 18. Needleweaving on a metal ring.

NEEDLE-WEAVING

In needle-weaving the weft threads are not all taken across from side to side, but woven in small groups on to warp threads using a needle. These warp threads may be moved across at random from one group to another and incorporated into another 'patch' perhaps with a variety of yarns. The ends are darned in afterwards, or as work proceeds. When you are needle-weaving on perspex, the warp threads are wrapped right round, being prevented from sliding about as they are pulled sideways by a strip of double-sided sticky tape placed across the top and bottom edges. Weaving can then take place on both sides of the perspex, and placed on top of a photograph of a house or landscape, to suggest three dimensions.

Needle-weaving on a ring makes an interesting mobile or window-decoration. Bind the ring with yarn first, and attach warp threads to this at angles, using a needle. (Fig. 18)

Fig. 19. Landscape, 5 in. × 5¼ in. (12.5 cm. × 13.5 cm.).

Fig. 20. Sunset sky, 5 in. × 4¼ in. (12.5 cm. × 10.5 cm.).

CARD-WRAPPING WITH YARN

These miniature landscapes are created by wrapping yarns around pieces of card, and framing them in a cut-out window: no stitches of any kind are needed. Only short lengths of yarn are required, although a more interesting and realistic result will be achieved if different textures are used. Positive colours and rougher textures are usually placed at the lower edge (this being the part which suggests the foreground) and paler, smoother yarns towards the top, with perhaps fluffy yarns for sky or surf.

Method

Choose a fairly small piece of narrow, thick card, and stick double-sided sticky tape down each side of the back (*a*). Begin to wrap yarn round the card, from the top, bottom or centre. Make sure that the threads lie neatly and closely together, leaving no gaps. Any ends which overlap on to other yarns instead of the sticky tape may be stuck down with glue. Change yarns frequently to create effects of changing light and the textures of land- or sea-scape.

Scenes which may be suggested by this method include:

Spring: pale blossom colours, pale greens and yellows.
Summer: brilliant colours to suggest fields of flowers.
Autumn: soft oranges, browns and reds, with golden yellow.
Winter: grey, silver, white, mauve and browns.
Seascape: pale blue-greys, white, sandy and 'pebbly' yarns.
Lakeside: watery colours, smooth and metallic yarns.

More wrapped card ideas may include knitting and crochet like those seen on pages 24 and 25.

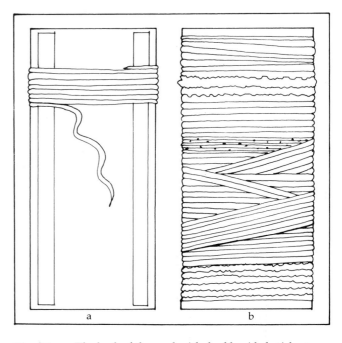

Fig. 21. a. The back of the card with double-sided sticky-tape in position, and wrapping begun.
b. The front of the card with wrapping completed, showing a mixture of textured yarns, some of which cross over at angles, but not too steeply.

16

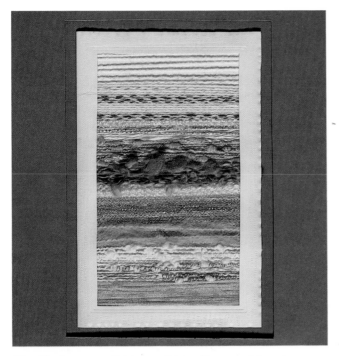

Fig. 22. Seascape, 3½ in. × 6¼ in. (9 cm. × 16 cm.).

Fig. 23. Snowscape, 4½ in. × 6 in. (11 cm. × 15 cm.).

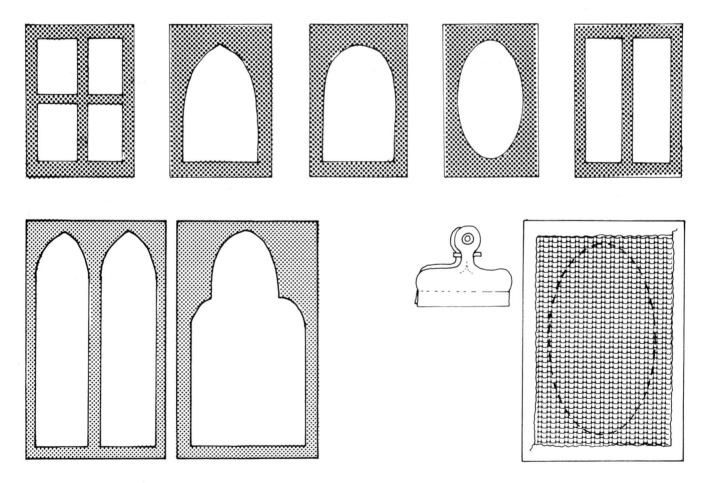

Fig. 24. Cut different shaped windows from thick coloured card to frame your wrapped landscape. These are some suggested shapes.

    Ensure that the frame entirely covers the wrapped card. Use a strong glue, with bull-dog clips to hold it all together until dry. The diagram shows the reverse of the card.

THE EARLIEST SPINNERS

In many countries, sheep have been bred for their wool for thousands of years. At one time, before there were any mechanical spinning machines, every woman in the house knew how to shear (or pluck) the sheep and how to prepare the wool for spinning.

The drawings on this page are taken from old manuscripts and show (top) a shepherd in Anglo-Saxon England shearing a sheep in an enclosure; (middle) a lady using metal clippers (or shears), while the lady in the lower drawing (left) is combing the raw fleece on sharp spikes which have been set into a log of wood. Carders, like the ones used by the lady in the centre, are still used by hand-spinners nowadays to comb the wool into long rolls called rolags which you see by her side. These are then taken up by the spinner who draws out the wool into a long thread and twists it on to a wooden drop-spindle to make yarn. The lady on the right uses a distaff tucked under her left arm on to which she has loosely tied a bundle of wool so that she has a continuous supply to hand. These methods are still used by some people today.

CHAPTER TWO

Start to knit

Tools – Knitting needles and crochet hooks – Experiments
notebook – Cast on and cast off –
Yarn/card wrapping with knitting – Basic stitches – Tension –
'The princess and the pea' – Dominoes and dice – Diagonal
squares – Cubes and boxes – More ideas for squares – Owls and
pussycats – Figures of fun – A mouse in the house – Making
shapes

*Alison is now going to begin knitting, but she must first
catch her yarn!*

Start to knit

TOOLS – KNITTING NEEDLES AND CROCHET HOOKS

One of the delights of using yarn is that the tools needed are few and simple, and easily carried around with the work. Originally the tools used by the early knitters and crocheters were handmade from wood or bone and did not conform to particular sizes as they do today. Now that we have patterns and yarns which call for specific instructions – which are relied upon by people everywhere to give them accurate results – the sizing of tools has had to be standardised. Nowadays, our tools are made from either metal, plastic or wood, with finer ones in either aluminium or steel. The very bendy plastic ones are not easy to use, but your choice will depend upon your own personal preference and involvement. Try out several types to see which you prefer.

Knitting needles are made in different thicknesses and lengths, but crochet hooks, though the thickness varies, stay more or less uniform in length.

The *thickness* will depend upon the type of yarn you are using. The *length* (i.e. short, medium, or long) will depend on how many stitches you are working with, although this does not apply to crochet. In metric sizing, the higher numbers indicate the thicker sizes. Old United Kingdom sizing is not used nowadays, but may still be found on older tools.

Apart from the usual knitting needles with knobs at one end, there are others too which are equally useful for everyday knitting.

The *circular needle* is a continuous length of plastic with a metal knitting point at each end. This is very useful for it may be used to knit tubular seamless pieces, flat circular pieces, or flat straight pieces with a large number of stitches. They can be bought in various lengths to accommodate any number of stitches. (You can never drop a needle when you are using a circular one!)

Double-pointed needles are bought in sets of four or six. They are used to knit circular or tubular pieces and are usually fairly short.

A *cable needle* is very short with a point at both ends. It is used for holding stitches at the back or front of the work when making cable patterns. They are made in two sizes, and some have a little bend in the middle to prevent stitches slipping off accidentally.

Crochet hooks tend to be more or less the same length, except for those used in Tunisian crochet, as the stitches in crochet are not held on the shaft of the tool in the same way as the knitted ones are. However, the sizes of crochet hooks and knitting needles do not exactly correspond. The number on the hook refers to the diameter of the stem, not the tip. Other crochet tools, such as the longer hook for Tunisian crochet, are fun to experiment with: the 'Cro-pin' with its point at one end and hook at the other is something else to be discovered in some shops. The U-shaped metal 'hairpin' is also useful for experiment as it is designed for making a lace-like strip of fabric; but it also is used for making a substantial cord.

Fig. 26. 'Hanging things', a detail from a wall-hanging made entirely of long strips of knitting, crochet, plaiting, French knitting and finger-cords, with tassels, pompons, bells and bobbles on the ends. Feathers and beads have also been added for extra effect. The colours all belong to one colour-family.

There are general rules about which size hook or needle to use with which types of yarn, although when you are experimenting you may try them all together to find out for yourself. With a thick hook or

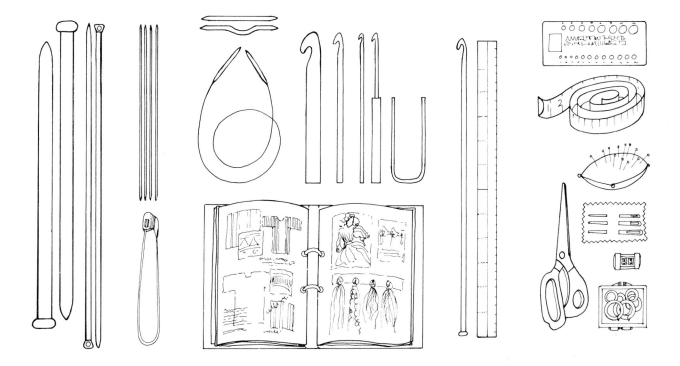

Fig. 27. The tools of knitting and crochet.

needle you would use a thick yarn, and with a fine hook or needle you would use a finer yarn. Much depends on the kind of fabric you want to make, and your own *tension*, that is, how tightly you hold the yarn and make stitches.

Here is a guide, but it will vary a little each way:
With 3-ply or 4-ply yarn you use a size 2¾ mm–3¼ mm needle, or size 2.50 mm – 3.50 mm crochet hook;
With double-knitting yarn you would use a size 3½ mm – 4½ mm needle, or a size 4.00 mm – 4.50 mm crochet hook;
With a thicker double-knitting yarn, such as Aran, you would use a size 5 mm – 5½ mm needle or a size 5 mm – 5.50 mm crochet hook;
Chunky yarns and bulky knits need very large needles like 7 mm – 9 mm, or a very large hook such as the 7.00 mm, 8.00 mm and 9.00 mm sizes.

Open, lacy fabrics can be made easily and quickly using the thickest hooks and needles, sometimes called Jumbo, Rocket or Maxi pins. Other useful tools and accessories are listed below:
Small scissors;
Roll-up needle/hook bag to keep tools tidy and safe;
Tape measure and/or ruler;
Blunt-ended yarn needles;
Stitch holders to keep stitches which are not being worked;
Knit tally: this fits on to the needle and keeps track of the number of rows worked;
Gauge for measuring the sizes of needles and hooks of which you are unsure.

Never use crochet hooks or knitting needles for any other purpose, for they may otherwise become scratched and bent. Keep them clean and in pairs, and well out of the way of small children!

THE EXPERIMENTS NOTEBOOK (Fig. 57, page 34)

Whilst not all experiments using yarn will mean the use of knitting needles or crochet hooks (there are many in this book which do not require either), you will find it useful to record in your notebook experiments in tension and density in both knitting and crochet. Not only will these little samples be an attractive record of something you discovered personally, but they will also be your reference when you want to know how many stitches to cast on to make a strip of knitting, say, 3 in. (8 cm) wide, what needles, which yarn, and how many rows will make it the required length. As people's tensions vary considerably (some people knit and crochet loosely and some make very tight stitches) this affects the number of stitches and rows they need to get the correct measurements.

Experiment with as many different yarn thicknesses as you can, and with as many different sized needles and hooks: a thick yarn used with fine needles produces a dense, hard fabric, while a fine yarn on thick needles or hook will produce an open, lace-like fabric.

Make a note in your book of the yarns you used, their names, where you obtained them, when, and the price. Stick a length of the yarn on to the page alongside your notes, and put the ball-band in too if you can.

Record which stitches you used in knitting and crochet, and what else you used the yarn for. Note the needle or hook size here too. Put in a drawing (or photograph) of what you made with the yarns, giving sizes and any other useful reminders.

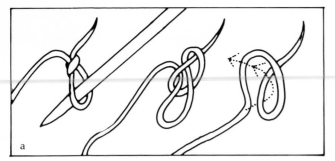

Fig. 28. Cast on.
a. Make a slip-knot and put it on to the left-hand needle.

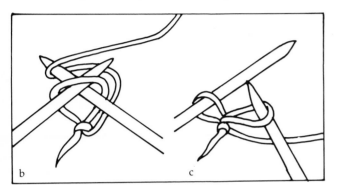

b. Put the right-hand needle into the slip-knot as shown, and take the yarn over the top and back again.

c. Pull the point of the right-hand needle through, with the loop on it.

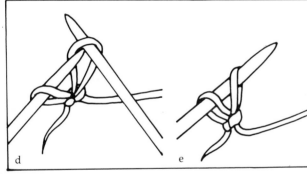

d. This loop is now transferred to the left needle.

e. Now there are two stitches.

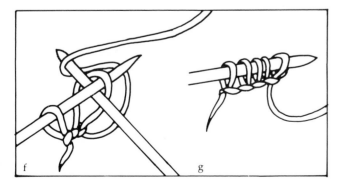

f. Enter the right-hand needle, this time between the two stitches, and throw the yarn over as before. Pull through again, and transfer the loop on to the left needle.

g. Repeat these moves to collect the stitches on to the left needle.

CAST ON AND CAST OFF

If you are knitting for the first time, you will find it easier to use a fairly thick needle and yarn, say a size 4 mm needle (old size 8) and a double-knitting (or thicker) type of yarn.

There are many different methods of casting on, all of which produce slightly different edges, some loopy, some firm and solid. The method shown here gives a neat edge for most purposes and is widely-used (Fig. 28).

To knit (Fig. 29)

a. The needle with the stitches on it is held in the left hand. Hold the yarn and the other needle in the right hand. Put the point of this needle into the first stitch, as shown, with the yarn at the *back* of the needle.
b. and *c.* Pass the yarn around the point of the right needle, and draw the loop through as shown in the diagram.
d. Now there is a new stitch on the right needle. Do this again with each of the other stitches until you reach the end of the row, and all the stitches are on the right-hand needle. To begin a new row, transfer this full needle to your left hand and use the empty one in the right as before.

To purl (Fig. 30)

a. The yarn is held at the *front* of the work, and the right-hand needle enters the front of the stitch as shown.
b. and *c.* Take the yarn over the point of the right needle and pull the loop through, dropping off the 'used' stitch (*d*).

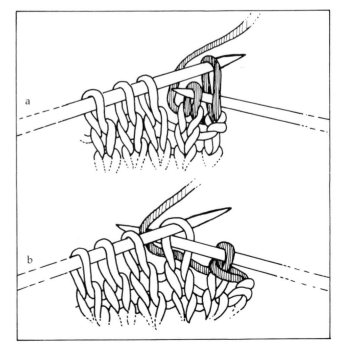

Fig. 31. Cast off.

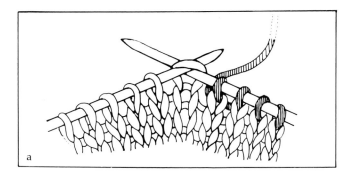

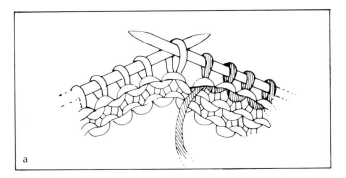

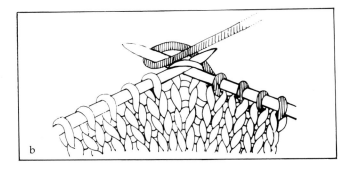

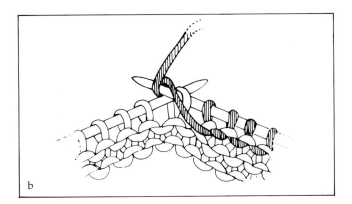

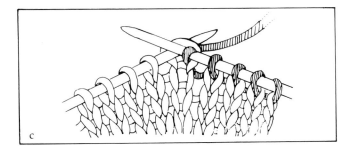

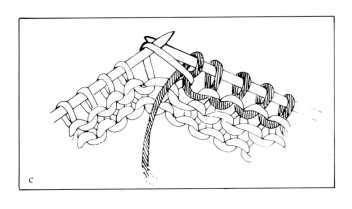

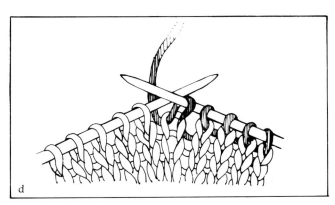

Fig. 29. The knit stitch.

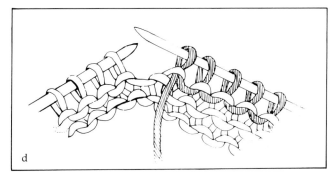

Fig. 30. The purl stitch.

Cast off (Fig. 31)

Knit (or purl) the first two stitches. Then, with the point of the left-hand needle, lift the first stitch over the top of the second and let it drop off the needle. One stitch remains on the right-hand needle. Now knit (or purl) another stitch, and repeat the process to cast off another stitch. Continue in this way until only one stitch remains on the right-hand needle, break off the

yarn and pass the end through the loop. Pull the end firmly and close the loop.

When you cast off on a purl row, simply purl each stitch instead of knitting it; and when you cast off in a pattern which combines both knit *and* purl stitches, remember to knit or purl the appropriate stitch before passing the previous one over it.

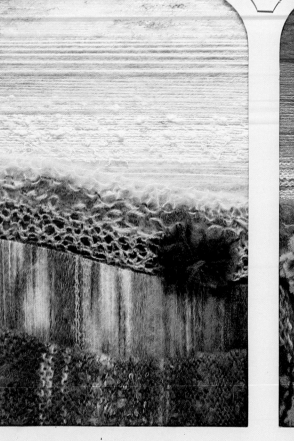

Fig. 32. a.b.c. and d: Dawn, noon, dusk and night.

a   b

YARN/CARD WRAPPING WITH KNITTING

A set of four panels – dawn, noon, dusk and night (Fig. 32 *a.b.c.* and *d.*) – each measuring 9¼ in. × 10½ in. (23.5 cm × 27 cm), depicts a continuous landscape seen at four different times in twenty-four hours. Each one is made in a combination of wrapping and knitting, with pompons here and there. The panels are seen through four windows, an idea which adapts well to the four seasons, or any other subject which deals with a progression.

Lines which travel from side to side are easiest to wrap in this way, although one narrow section of card has been wrapped the other way up for a different effect. The sky was wrapped as far down as the horizon, then a piece of rough knitting in Garter stitch was placed next to suggest distant trees. The narrow strip of wrapped card in the centre overlaps this piece, and this in turn is fitted against the lower piece of card, which is covered with more rough knitting. Crochet could be used equally well instead of, or as well as knitting. Pompon trees not only add to the three-dimensional effect, but will also serve to cover up areas that need hiding!

The moon in the night scene is a circle of crochet stuck on top of the threads, and here and there a random-dyed bouclé yarn was laid across the top. This explains how the white areas occur in the same place underneath the moon.

The two lower pieces of card fit next to each other and are cut from one piece. The lower one is covered with knitting and the other with wrapping (Fig. 33*a*). All pieces fit on to the large background piece.

The knitting is pulled over the shaped card and laced in position across the back (Fig. 33*b*).

Each piece is then stuck down with glue, and the whole picture framed with a card 'window'.

This combination of wrapping, knitting and crochet may be especially useful to teachers, where some pupils in the class may be able to produce rough pieces of knitting, while others can wrap neatly or make pompons. The shapes of the knitting/crochet pieces are not critical as long as they cover the card at the bottom and will stretch from one side to the other for the higher ones. Open-textured pieces (even with

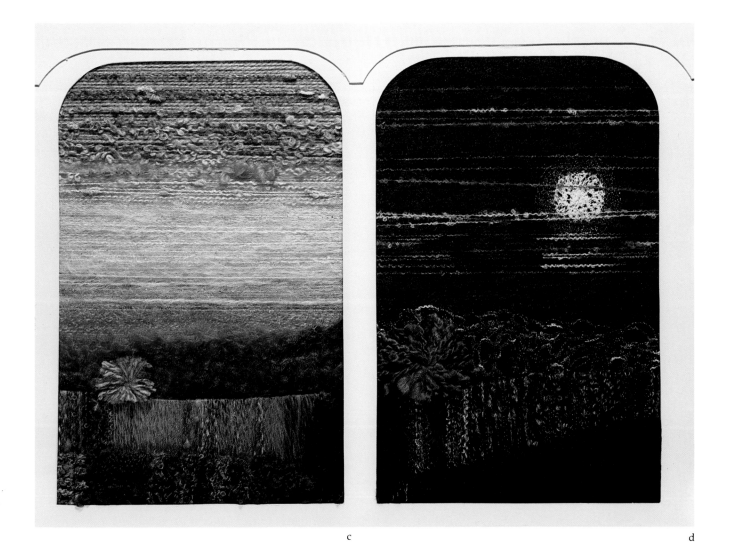

c

d

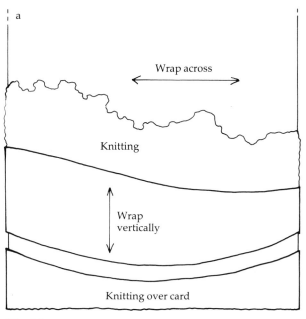

Wrap across

Knitting

Wrap
vertically

Knitting over card

Fig. 33.

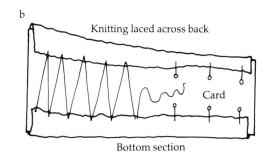

Knitting laced across back

Card

Bottom section

holes and other 'mistakes') are just as useful as neatly
made ones, while those yarns which are more difficult
to knit or crochet, such as slubs or mohair, can be used
for the wrapping. This also uses much less yarn than
either knitting or crochet, even though half the yarn
used is hidden on the back. It is important to use card
of a matching colour to frame the work.

25

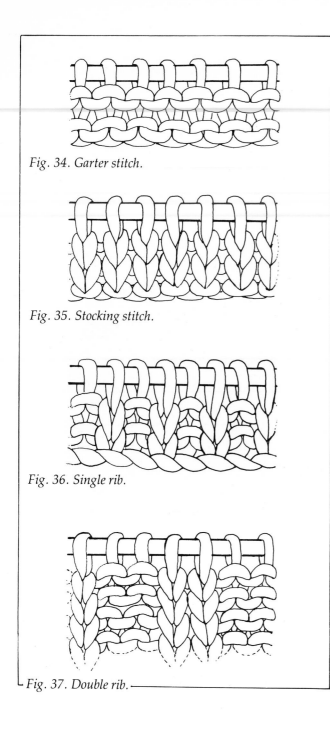

Fig. 34. Garter stitch.

Fig. 35. Stocking stitch.

Fig. 36. Single rib.

Fig. 37. Double rib.

MISS THICKENTHIN   MISS LOOSENTIGHT

Fig. 38. Miss Thickenthin is knitted on one pair of needles, but with yarns of three different thicknesses. Miss Loosentight is knitted with one yarn, but with three pairs of different-sized needles.

BASIC STITCHES

Garter stitch (g.s.). Every row is knitted.
Stocking stitch (s.s.). One row knit (k), one row purl (p), alternately.
Single rib. On an even number of stitches, k.1, p.1, alternately.
Double rib. On a multiple of four stitches, k.2, p.2, alternately.

Many patterns are made by combining the knit and purl stitch in various ways. Moss stitch, for instance, is made like a single rib on an *odd* number of stitches and beginning every row with a k.1. Then every knit stitch will fall above the purl stitch in the previous row. Double Moss stitch works in the same way using two k. stitches (sts) and two p. sts.

TENSION

As you will see by looking at Miss Thickenthin and Miss Loosentight (Fig. 38), strange things can happen when different yarns and different tools are used. Imagine all the effects which could be produced by using the whole range of needle sizes and all the different yarns you could find. The permutations would be endless!

This is why tension plays an important part in the success or failure of our efforts, even one size thicker or thinner (needle *or* yarn) can make a considerable difference in the overall shape of a piece. Children will enjoy finding this out for themselves, and the small decorative projects which follow are a good way of making this fun, as nothing is too large, or takes too long, or needs to fit! In any case adults can usually find a way of adapting a piece which turns out too large, or too small, by making pleats and gathers, or by adding a few rows of crochet.

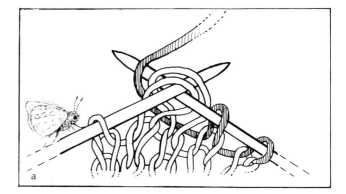

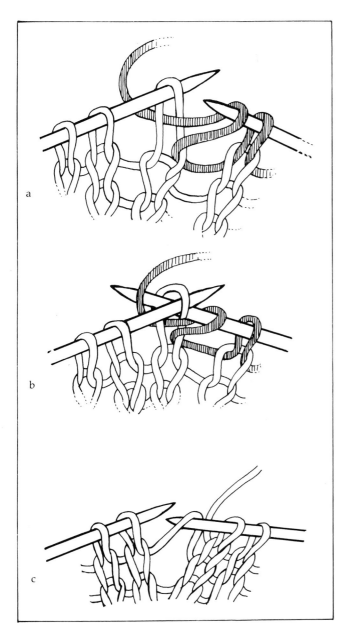

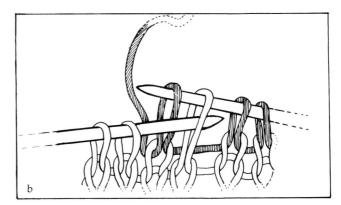

Fig. 40. Decrease:
a. Knit two stitches together (k.2tog.); or
b. slip one stitch, without knitting it, on to the right-hand
needle (this is a 'slip stitch' or s.s.) then k.1, and pass the slip
stitch over the top of the knitted one (called p.s.s.o).
Try these two decreases out for yourself to discover the
different effects they produce.

Fig. 39. Increase:
a. Knit the stitch in the usual way, but do not let the 'used'
stitch slip off the needle;
b. Instead, swing the point of the needle round to the back of
the stitch and knit into it again, making two stitches out of
one;
c. Another way of increasing is called 'Make One' (M.1).
With the point of the right-hand needle, pick up the thread
which lies between two stitches and knit into it as though it
was a stitch. This will make a decorative hole in the fabric, so
if you do not want a hole you can move the 'picked up' thread
over on to the left-hand needle first and then knit into the
back of it.

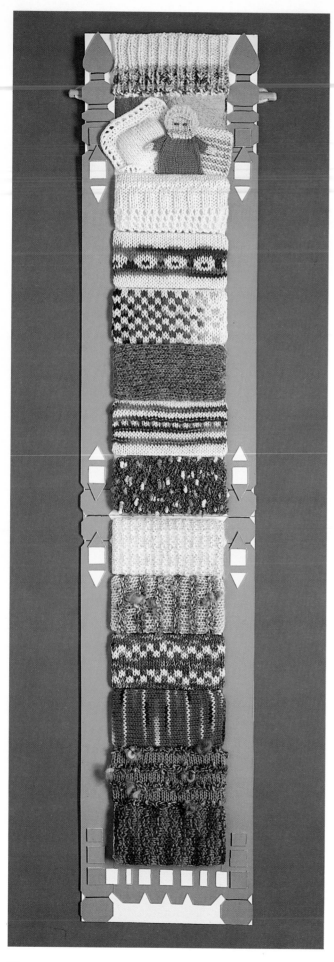

Fig. 42. Detail showing the very bottom mattress, which hinges to reveal the pea which the Princess found so uncomfortable.

THE PRINCESS AND THE PEA (Fig. 41)

The only way for the queen to prove that the young girl was really a princess, as she claimed to be, was to put a tiny pea underneath twelve mattresses. If she was kept awake all night in discomfort, then she must truly be a princess and would be asked to marry the prince. Of course, we all know that she *was* a real princess, – and didn't sleep a wink!

This delightful wall-hanging is made entirely of oblongs and squares of knitting, worked from the wide or narrow edges in any simple stitch and yarn. The dimensions need not be too exact either, as pieces which are too small may have rows of crochet added (or an edge all round), and too large pieces may be turned under before they are stapled down.

Each oblong of knitting is stretched over a piece of card, with a layer of Terylene padding sandwiched between the two. Lace the knitting gently across the back of the card to keep it in place and then arrange the twelve mattresses on to a larger piece, and stick them down with glue. Cover each join with pieces of ribbon, lace, cords or crochet.

The princess, also, is made from tiny oblongs of knitting stuck on to card, with hair made from a strip of crochet. Her cushions are made in the same way as the mattresses, with the valance at the top.

The pea is sewn on to a 'false mattress' at the very bottom, which is hidden by another loose mattress which is hinged to the one above, as seen in Fig. 42. A tiny piece of Velcro will keep it closed.

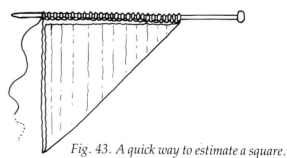

Fig. 43. A quick way to estimate a square.

Left. Fig. 41. 'The princess and the pea'. 40 in. × 8¼ in. (101.5 cm. × 21 cm.). Each mattress measures 2½ in. × 5¼ in. (6.5 cm. × 13.5 cm.).

Fig. 44. Dominoes and dice.

DOMINOES AND DICE

Simple oblongs and squares can also produce three-dimensional ideas like these colourful dominoes and cubes. Use empty cardboard boxes for the cubes, or construct one from six card squares. Wooden cubes made from off-cuts will make firm and solid shapes and would be especially suitable for paper-weights, children's building bricks and door-stops. The dominoes are an interesting variation on the traditional game and can be useful for handicapped and very young people as the colours and stitch-textures provide the necessary information instead of dots.

The stitch-textures are all made by combining the knit and purl stitches; no shaping is required. The back of each piece is plain black stocking stitch, and a piece of card acts as a stiffening between the two layers.

Smooth yarns produce more accurate shapes than highly-textured or fluffy ones, so this project is useful for using up left-over oddments of double-knitting or 3-ply yarns. Also, the rows will be easier to count, and the pieces easier to sew up, giving a neater finish. Make sure that the yarns you use on each piece are of the same thickness – otherwise one part may be bigger than another. Keep a note of how many stitches, which needles, and how many rows you needed to make a square or oblong.

There is a quick way of estimating whether you have worked enough rows. This is by folding the work into a triangle as shown in Fig. 43. When the side edge measures the same as the cast-on edge, your knitting should be about square, but do not stretch the fabric in order to ensure this.

The cube

There are two methods of knitting the squares which makes the sides of the cube. They can be knitted all in one piece (Fig. 48a, overleaf) or in six separate squares which are joined together afterwards. The all-in-one method is quick and useful when only one yarn is used, and when making a box with a hinged lid. The lining and the card stiffening can then be made to exactly the same pattern and measurements.

The box (overleaf)

One long strip of knitting was made to cover the four sides; this was then sewn up into a tube. Each piece of card was enclosed in a 'pocket' of knitting made up of the outside and lining, and the pockets were then sewn together to enclose each piece of card in a square formation. The bottom and top of the box are made from square pieces; the base lining is four triangles of different colours, sewn together.

Fig. 45. Make a cube into a cottage.

Fig. 46. Make one in a mixture of geometric patterns.

Fig. 47. A knitted Fair Isle box.

MORE IDEAS FOR SQUARES

1. Knit square or cubed food such as biscuits, liquorice, sweets and toffees. (For humbugs, see page 48.)
2. Knit a box inside a box, inside a box – each one smaller than the last.
3. Make a patchwork square stitch-sampler, either to hang on the wall or as a functional item.
4. Fold them into triangles with a stiff piece of card inside and hang them as mobiles decorated with tassels and pompons, bells and bunches of feathers.
5. Make a Jack-in-the-box with a glove-puppet inside – made from squares of course!

Diagonal squares

The black, white and grey cube is knitted in diagonal squares (see also the patchwork hanging on page 63), which produce a totally different appearance when put together in this way. The diagram (Fig. 48b) will explain how the colours are arranged so that they join on the corners.

To knit a diagonal square, begin by casting on 2 stitches.

Working in Garter stitch (this is important) increase one stitch at each end of *every other* row until there are 30 stitches on the needle. Knit one more row and then begin to decrease in the same way until 2 stitches remain. Cast off.

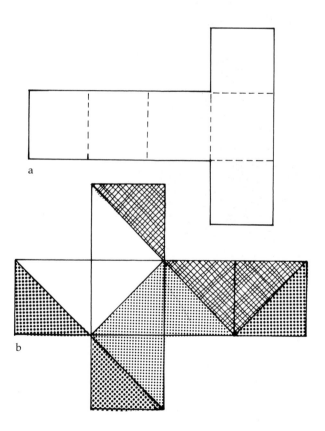

Fig. 48. a. The all-in-one cube pattern.
b. The diagonal cube pattern. Six squares, knitted on the diagonal, with the colours arranged as shown, will produce a cube like the black, grey and white one seen in the colour photograph.

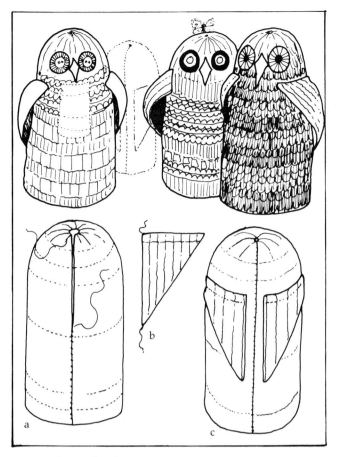

Fig. 49. Knitted owls.

Fig. 50. Knitted cats, the main body and the face.

OWLS AND PUSSYCATS

These endearing and ever-popular creatures are made very simply from one basic piece of knitting (or crochet, if you prefer). The owls have two small triangular wings added, and the cats have sewn-on tails. While the cats require no shaping at all, the owls only have a little decreasing on the last few rows, and on the wings. The base of each creature is a cardboard disc which helps them to stand firmly. They are about 3½ in. (9 cm) high, although this will depend on which yarn is used.

The owl

Both the right and reverse sides of the fabric may be used.

With a double-knitting yarn and 4.00 mm needles, cast on 30 stitches.

Knit 30 rows, then shape the top by knitting two sts. tog. all across the row to make 15 sts.

Knit one more row, then break off the yarn, leaving about 8 in. (20 cm) to thread through the sts. Draw these up tightly and sew the back seam.

The two small wings are made on 8 sts. Work 2 rows, then k.2tog. at the beg. of every other row until 2 sts. remain. Cast off, and make another wing to match. The diagram shows where these should be sewn to the body. Embroider the face, using buttons for eyes, or felt. Stuff the body cavity, (clean, chopped-up nylon tights make a good washable filling) and glue a disc of card inside the base.

The cat

You can make different sized cats by using yarns of varying thicknesses and by using thick and thin needles on the same number of stitches. The body is a rectangle of knitting on 30 sts. working for 32 rows (any stitch you like). Join the rectangle to make a tube with the seam down the back, then sew across the top of the head.

Now sew across each corner to make the ears, as shown in the diagram. Embroider the face before padding, and add whiskers. Pad the body cavity, then make the tail on 4 sts., which should be about 3 in. (8 cm) long. Sew it in place on the centre-back seam, and curve it around the body – it can be glued or sewn in place. Insert the disc of card and glue this in place to make a firm base.

You may make these owls and pussycats into egg-cosies by leaving out the padding and the card base.

FIGURES OF FUN (Fig. 51, overleaf)

A colourful mobile can give hours of happiness, especially to a small child or someone confined to bed. These little figures are made of small rectangular pieces stiffened with card but, like the owls and cats, could also be used as cosies or as puppets for small hands. No shaping is needed for the main pieces, and very few stitches are required. Only some of the hats need shaping, and St Francis's butterfly!

The size of your figures of fun will depend on the kind of yarn and needles you use, but the ones you see

Fig. 51. Figures of fun – a seven-part mobile of knitted characters.

here were made with double-knitting yarns on size 4.00 mm needles. You will need about 16 sts., and the diagram suggests the number of rows to work for the face and body. Stiffen each 'pocket' of knitting with thick card (*c*) measuring about 2½ in. × 4½ in. (6 cm × 11 cm) and finish off with embroidered faces and extra hair if needed. The eyes have been cut out of felt and stuck on with glue.

St Francis's butterfly is made of four tiny triangles sewn together and then fastened to a metal ring to look like a shiny halo. Other small additions make quite a difference to the finished effect, including beads, bells, feathers, buttons, cords, ribbon and lace. Use combinations of knit and purl stitches to vary the texture and pattern of each figure, as well as using different coloured and textured yarns.

Groups of figures relating to a story or theme could include:
The Nativity
Your family
Favourite story characters
Clowns and circus people
Nursery-rhyme characters
St Francis and other saints
Fairies, elves and gnomes
Water spirits and other unearthly creatures
Magicians and witches
Robots and spacemen
Pied Piper and children, burghers and rats
Snowmen, Santa Claus, etc.
Ballet dancers (complete with tutus)
Snow White and the Seven Dwarfs

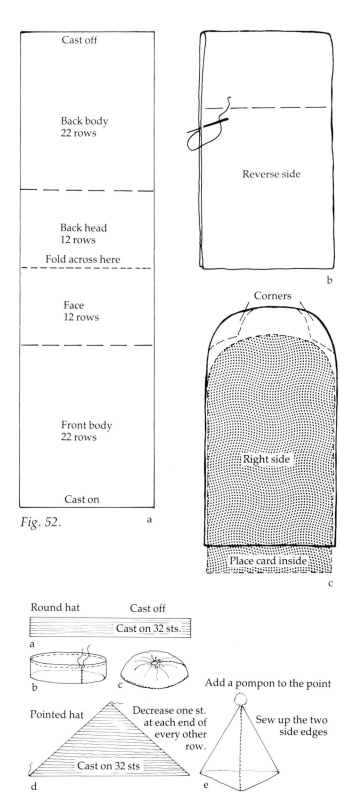

Fig. 52.

Round hat   Cast off

Cast on 32 sts.

a

b   c

Add a pompon to the point

Pointed hat   Decrease one st.
at each end of
every other
row.   Sew up the two
side edges

Cast on 32 sts

d   e

Fig. 53. Hats for figures of fun.

A MOUSE IN THE HOUSE

I did not dare to place this mouse too close to page 31, but here he will be perfectly safe. He can be made from very small oddments of double-knitting yarn and a pair of 3.50 mm needles. (If you use finer yarn and needles, your mouse will be smaller even with the same number of stitches.) Use grey yarn for an ordinary house-mouse, white (with pink eyes) for a pet one, and pink for a sugar one.

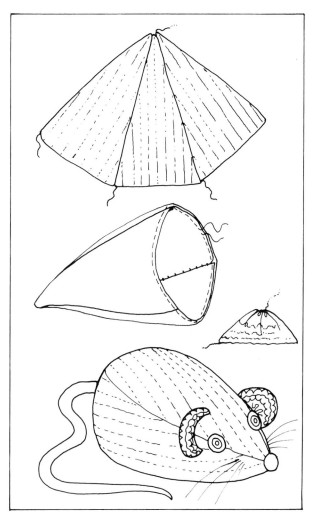

Fig. 54.

Instructions

1. He is made in three sections, and a tail
2. Use D-K yarn and 3.50 mm (or 4.00 mm) needles cast on 14 sts.
3. Work in s.s. for 4 rows.
4. Next row; k.2tog at both ends of the row to make 12 sts.
5. Next row; purl
6. Next row; knit
7. Next row; purl
8. Repeat the last 4 rows until there are only 4 sts. left, then k.2tog. twice. Purl one more row on these 2 sts.
9. K.2tog. and finish off.
10. Work two more pieces the same and sew these together as shown.
11. Run a gathering thread around the opening, place padding into the cavity and draw the thread up, securing with a few stitches. Make a crochet chain, or a cord, for the tail. Attach this with the same thread.
12. The ears are made on 10 sts., knit 1 row, then k.2tog. 5 times to make 5 sts. Gather these sts. on to a thread and fasten off. Make two.
13. Attach the ears to the head on a slight curve; embroider the nose and eyes (buttons will do nicely) and attach a few long whiskers of bristle.

MAKING SHAPES

Once you know how to knit, purl, increase and decrease, then you will easily be able to make simple shapes which can be used as 'building blocks' to compose more complex shapes. The ones shown here all use increase and decrease added to squares and oblongs, and are useful for making buildings, abstract designs, figures, flowers and animals, aerial views of gardens, trains and many other subjects.

When you have decided on the shapes that you need, choose:

1. The colour and type of yarn
2. The needle size
3. The stitch pattern (e.g. Garter st., Stocking st.)
4. The number of sts to cast on

Always make a note of these experiments in your notebook to remind you of what you did and what to expect next time.

You may need to press (or 'block') your shapes before you use them: if so, pin them down on to an ironing board with the pin-heads lying flat so that the iron cannot push them into the board. Press them *very gently* over a damp cloth: they will not need to take the full weight of the iron. Now place the pieces on a background of fabric or strong paper, and keep them in position with pins until you are ready to sew or stick them down. Use an ordinary sewing cotton for this, in a matching colour.

Details like stems of flowers, door knobs, window-frames, lines and dots, can be embroidered on later, as can also buttons, beads and sequins. Ideas for knitted pictures made from small shapes include figures in a queue, masks, fish, starfish and pebbles, butterflies and flowers, patchwork fields and gardens, Noah's Ark and circus characters.

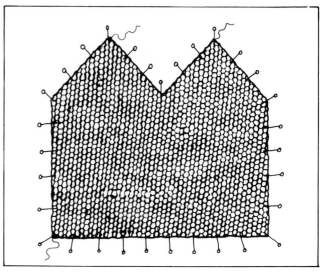

Fig. 56. Blocking a shape..

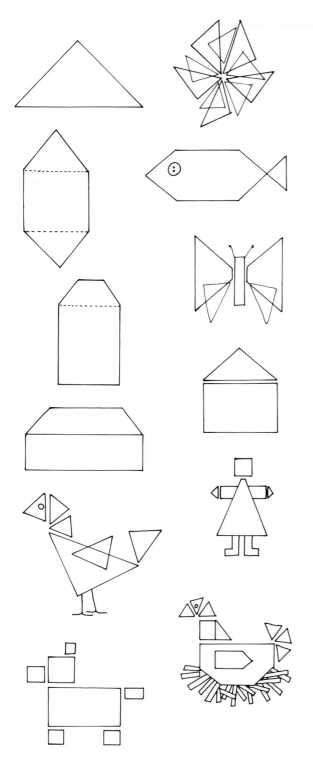

Fig. 55. Shapes using increase and decrease.

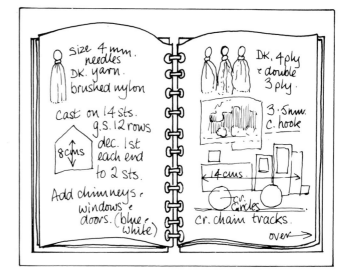

Fig. 57. The experiments notebook. See page 21..

CHAPTER THREE

Add some crochet

First steps in crochet – Chain – Method of making the crochet
stitch – Colourful cones and tubes – Knit into crochet and
crochet into knitting – Crochet in the round – Fantastic fungi –
Part-row knitting – Crochet in the square – Crispin, the crochet
cobra – Humbug! – The cacti collection – Gruesome glove
puppets – Surface chaining

*Pauline is crocheting a bag in which she will keep her
rainbow-coloured reflections.*

Add some crochet

FIRST STEPS IN CROCHET

These pages are especially designed for *complete* beginners in crochet. If you already know how to make a chain, and the basic stitches, then you can pass on to some of the projects on the following pages.

CHAIN

Fig. 59 on the left shows you how to make a chain, Use a fairly large hook, holding it as shown in *a*. Choose a yarn which is fairly thick and smooth, as fluffy and textured ones are not easy for beginners to manage. A size 4 or 4.50 mm, hook and D.K. yarns are most suitable to practise with.

You will notice that modern hooks have a flat part on the shaft with the size number on it. If you take hold of the hook with your finger and thumb on the flat part, the hook will automatically be in the 'turned down' position.

Once you have made a chain, you will understand how the hook passes underneath the yarn (called 'yarn over' – y.o.) to pull it through the loop, and the chain is the foundation of fabric-making in crochet. Howeverer, there is the difficulty, when practising stitches for the first time, of finding the correct hole for the hook to enter, keeping the chain still while you find it, etc. So on page 37 you can see a specially devised way of learning how to make the basic crochet stitch without having to wrestle with a chain at the same time. This method is ideal for teaching young people to crochet.

METHOD OF MAKING THE CROCHET STITCH

Take a strip of firm, unfrayable fabric like a heavy-duty interfacing (pelmet-weight Vilene is ideal) and punch holes across the top edge spaced about ¼ in. (0.5 cm) apart. Now choose a crochet hook which will pass through the holes comfortably, and a D.K. yarn, the end of which should be stapled firmly to the reverse side of the fabric to stop it from slipping about.

Your crochet stitches are worked from right to left, as in knitting, and *every* row begins with one, or more, chains. This is called the 'turning chain' and counts as the first stitch of every row; so it must be included in your counting.

Follow the diagrams to make the first basic crochet stitch, called 'double crochet' (d.c.). Five stitches are made, although one of these is the turning chain. Practise turning, making the first chain, then the stitches, counting five at the end of every row. When you have mastered these movements, you will now be able to crochet! It is only a simple step to making these same movements into a foundation chain.

Note: The hook is entered into the top stitch of each previous row, which is called a 'chain space' (ch.sp), and goes under *both* strands of the chain, not just one. There are various other stitches in which the hook enters different parts of the underneath stitch, but for basic stitches take it into the complete chain space.

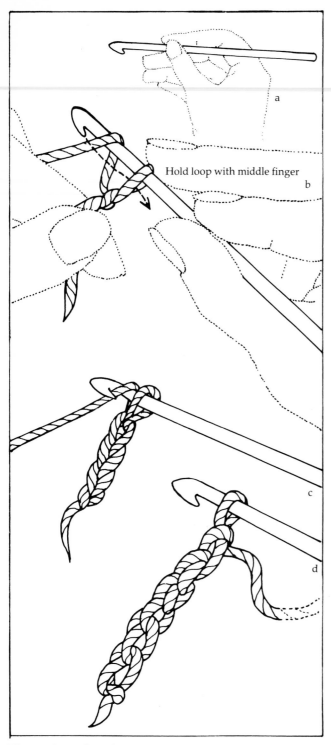

Fig. 59. To make a chain.
a. Hold the hook so that the end curves downwards like this.
b. Put the hook through the slip knot (see page 22 for this). Hold the tail-end of the yarn and the supply yarn in the same (left) hand as shown, and the hook in the other hand. Now place the hook under the yarn (this is called a 'yarn over') and pull it through the loop of the slip-knot. Continue to do this with your middle finger on the loop to stop it slipping off while you 'yarn over'.
c. This produces a chain. Here you see the flat side of the chain, which is the right side.
d. This is the other side of the chain – the round side.

Hold loop with middle finger

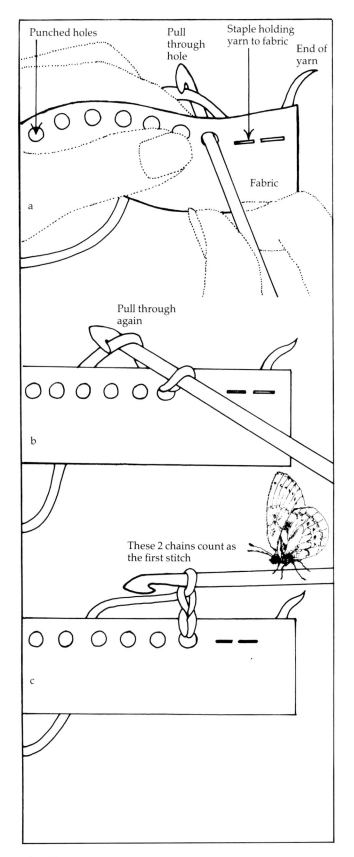

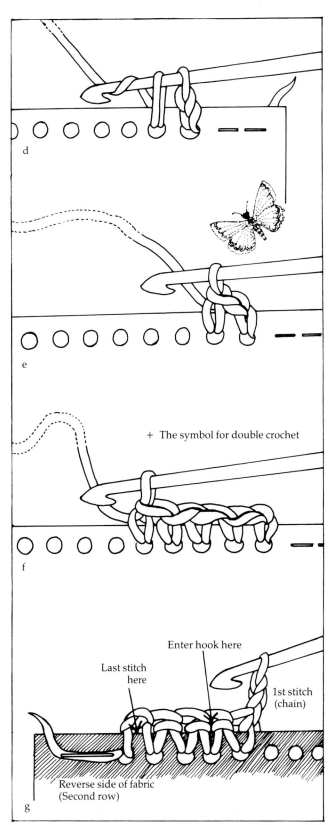

Fig. 60.
a. Enter the hook, yarn over (y.o.) and pull through the hole
b. Y.o. again and pull through the loop to make one chain.
c. Repeat b. to make 2 chains. This counts as the first stitch.

d. Hook through second hole, y.o. and pull through, y.o. again.
e. Pull through, stitch completed.
f. Repeat d. and e. in 3rd, 4th and 5th holes.
g. Turn the fabric round, make 2ch.s to begin the 2nd row, then repeat d. and e. entering the hook as shown underneath both strands of the chain space. Make 4 sts (5 including the first ch.), then turn again and begin the 3rd row.

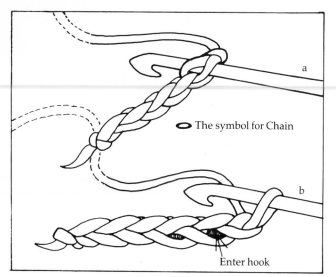

Fig. 61. The basic stitches
a. Showing how the hook is entered into the flat side of the chain, with two strands on top of the hook and one underneath.
b. For a double crochet (d.c.) the hook is entered into the second chain space from the hook (the first chain counts as the first stitch). Working into each chain space, complete the d.c. as shown in Fig. 60 d and e.

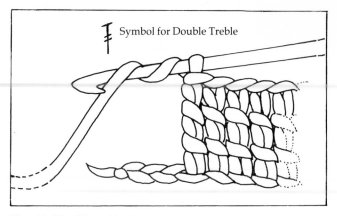

Fig. 64. Double treble (d.tr.). A very tall stitch, made by yarning over **twice** before entering the hook to make the first part of the st. As follows:
a. Y.o. twice, enter hook, y.o., pull through (4 loops on hook).
b. Y.o. and pull through 2 loops (3 loops left).
c. Y.o. and pull through 2 more loops (2 loops left).
d. Y.o. and pull through the last 2 loops.

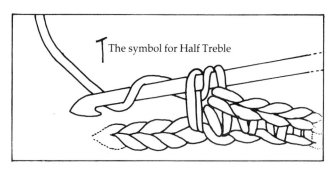

Fig. 62. Half Treble (h.tr.). Y.o. the hook **before** entering the chain space, then pull through. This makes three loops on the hook.* Y.o. again and pull through all these 3 loops tog. This is a taller st. than a d.c. and therefore needs a taller turning chain at the beg. of each row (2 ch).

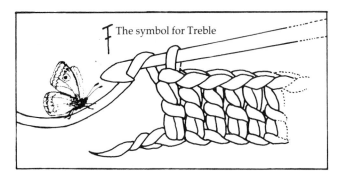

Fig. 63. Treble (tr.). This is an even taller st., and the most commonly used. Begin as for the h.tr. as far as the *. Then y.o. and pull through the 1st 2 loops, y.o. again, and pull through the last 2.

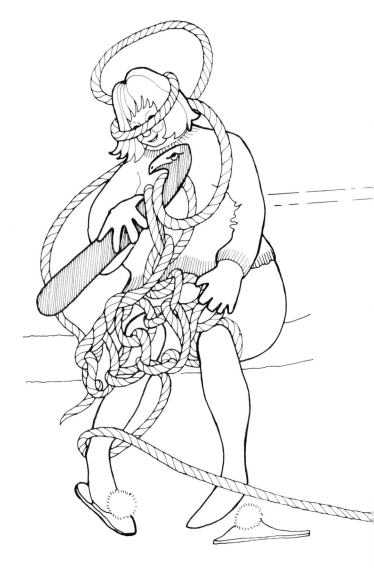

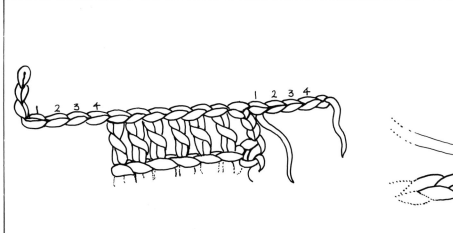

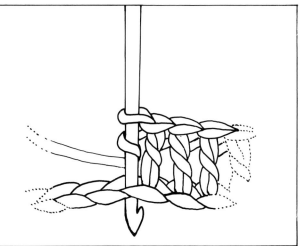

Fig. 65. Increase.
To extend the crochet outwards on one, or both sides, make extra chains with spare yarn on one side (right of diagram), then work across the main part of the crochet, working more chains on the other side before adding the turning chain and working back across all stitches. If only one side is needed, simply extend the chain as seen on the left side.

Fig. 66. To increase, make two, or more, stitches into the same chain space.

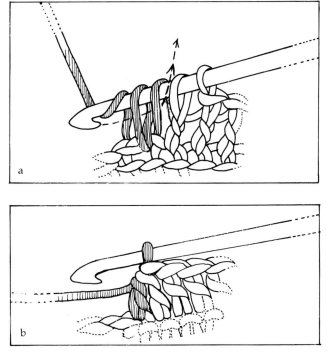

Fig. 67.
To decrease. This is done by making two unfinished stitches and then completing both together on the last pull-through.
1. Make a tr. into one ch.sp. as far as the last pull-through. Leave the last 2 loops on the hook.
2. Y.o. and begin another tr. into the next ch. sp. (this is the shaded stitch in the diagram).
3. The last part of the stitch is pulled through both the first and second trebles, pulling two sts. into one (b.).
The other way of decreasing is to miss a ch.sp., but this method leaves a hole. It is useful, though, for small children to understand.

Fig. 68. Cone heads.

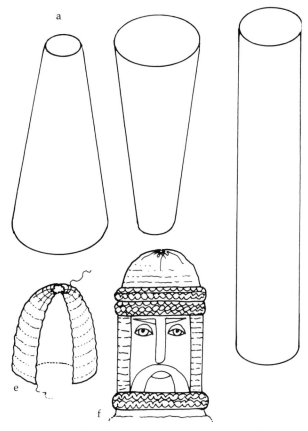

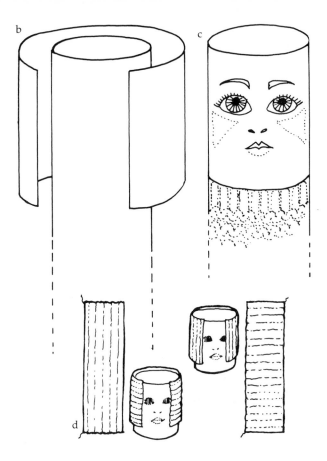

Fig. 69. Cone heads

a. Knitting machine yarn cones, or plain cardboard tubes used as bases for the heads and complete figures.

b. The paper on which the face is drawn is wrapped around the tube and joined down the back with glue.

c. The knitted or crocheted fabric covers the bottom edge of the paper face.

d. The side and back hair can be made in one long strip in either of these two directions.

e. For hair, or head-dress, gather one side of the knitted strip to curve around the crown of the head. Place knitted/ crocheted strips around the brow.

f. Without the tube, these figures can become glove-puppets.

Fig. 70. Tube figures.

COLOURFUL CONES AND TUBES

Crochet and knitting have been mixed together on the rectangles which cover these cones and tubes. Make your favourite storybook characters, spacemen, tall or short people, plain or fancy, fact or fiction. Use either a knitting-yarn cone or a card tube from the kitchen to make a head or a complete figure. There is very little shaping involved, you can use knitting or crochet, or both together. No arms or legs are indicated, and they can also be made into glove puppets by putting a very short tube into the head-part only, leaving the body free to take a small hand instead.

Use as many different textured yarns as possible and vary the stitches too by changing from crochet to knitting and back again. (You can find out how to do this overleaf.) You will need cardboard tubes, cones, glue, blunt wool needles, paper clips (to hold the paper face-on while the glue dries) and coloured pens. Also beads, sequins, and rings etc.

The full-length tube figures are, from left to right:
1. Burmese lady, knitted in a simple lace stitch. Her cone hat is part knitting, part crochet.
2. Old lady in frilly hat, knitted in smooth and bumpy yarns. Her hat is crocheted, and folds over at the top to hang down her back ending with a very large tassel.
3. Tall lady with earrings: knitted and crocheted in smooth and bumpy yarns, with a crochet frill added. Her hair is a pompon of furry and glitter yarns, with an extra piece of knitting to cover the back and sides.
4. Little blue boy is made entirely of knitting.

5. Yellow spaceman: knitted lengthways and twisted round on the tube before glueing. The neck frill and hat are of crochet, the top is stiffened with card.
6. Red Indian girl: patterned crochet with a fringe down one side. Her black hair was made from a coil of plaited yarn.
7. Opera singer: her arms are made of one long plait threaded between the tube and the knitting, in at one side and out again at the other.
8. Little blue-haired boy: his knitted body has been padded, and his collar is crocheted.
9. and 10. Two little Russian dolls, one knitted and one crocheted. Made in one piece with a large hole for the face. The stitches were gathered at the tops, not cast off, and finished with a small tassel.

41

Hints on making the tube figures

1. *The tallest* figure in Fig. 70 (page 41) measures about 12 in. (30 cm) tall, and the shortest about 2 in. (5.5 cm). The number of stitches needed will depend on the thickness of the yarns and the size of the needles, but as a rough guide you will need about 28–30 sts if you are using D.K. yarn on no. 4 mm needles to go around a tube of about 2¼ in. (6 cm) diameter.
2. The head takes up about one-fifth to one-sixth of the total body length.
3. Sew the rectangle into a tube of knitted/crocheted fabric, and slide it on to the card tube until it reaches the neck.
4. Glue it in place at the top and bottom edges, on top of the paper face as shown on page 40. A face from a magazine may be used instead of a drawn one. Hair and hats can be made as shown in the diagrams or in any other ways using tassels, plaits, pompons and braids.

KNIT INTO CROCHET AND CROCHET INTO KNITTING

On many of the projects in this book, knitting and crochet have been used together to make the same piece of fabric. This is easy and useful to do, as you will find that some effects are easier to achieve in crochet than in knitting, and vice-versa. You will discover other advantages of changing techniques as you progress.

Fig.. 71 shows how this is done, by casting off knitted stitches and then crocheting into the cast-off edge with the same, or a different yarn.

Important note: crochet stitches are fatter than knitted stitches and so take up more room sideways. You must therefore compensate for this in any of the following ways;

a. Put fewer crochet stitches into the cast-off edge. This means that you will only need about two-thirds of the number of knit stitches, that is, two crochet stitches into every three spaces.

b. Use a finer crochet hook than the knitting needle used.

3. Use a finer yarn when you change from knitting to crochet.

If you want to change from crochet to knitting, your work may gather up more tightly unless you remember to change to either a thicker yarn, or thicker needles, or increase the number of stitches.

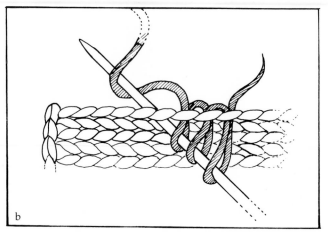

Fig. 71.
a. Crochet into the side-edge (or the cast-off edge) of knitting. The x marks the spot where the hook will enter next.
b. Picking up stitches with a knitting needle from a side-edge, or from a crochet edge.

CROCHET IN THE ROUND

In crochet, it is just as easy to work in rounds as it is to work in straight rows. The technique is the same, each 'round' beginning with a turning chain, each stitch made into the chain-space below. From the same beginning can be made discs, rings, cups, domes, tubes, spirals and a host of other round shapes, as wide or as narrow as you wish. In each case, the foundation chain (that is, the chain into which the first row of stitches is worked) is made into a ring by joining the last chain to the first with a slip-stitch. (See the Glossary on page 126 for explanation.)

Note. When you work in circles, rows are now called 'rounds'.

To make a flat disc, begin by making a chain of 6. Link the last stitch to the first stitch with a slip-stitch (*a*) to make a ring, then make two more chains to begin the first round – as you would if you were working in rows. Keep your left finger and thumb in the hole to keep it open and work about 12–14 trebles into it (*Note:* not the chain spaces). This will bring you right back again to the first stitch, and, again, you will link the last stitch of that round to the first stitch with a slip-stitch. Work two more turning chains ready for the second round.

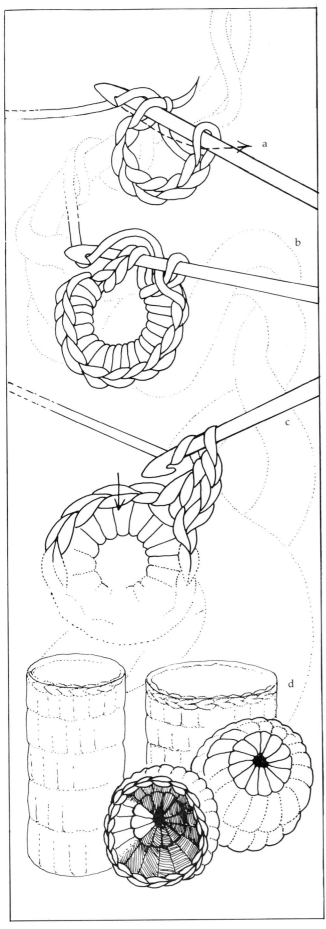

Fig. 72. Crochet in the round.

Now make two trebles (2 tr.) into each chain space (ch.sp.) all the way round (*e*). If you want to make the disc larger, you must continue to add more rounds, increasing as you go (because the circumference gets bigger!) but you will not need to increase on future rounds as much as you did on the first one.

To make a dome, or cup shape, the first round is the same as above, but on the second round work only *one* treble into each chain space instead of two. The edges will then curl up because there are not enough stitches to make it lie flat. Practise using these with different yarns and more or less stitches.

Making a tube is very similar, except that you will begin with more chains on the foundation ring. As you want to keep the hole open at one end, you will work the first round into each *chain space* instead of the centre ring, without increasing, i.e. unless you want a thick and thin tube.

To work in a spiral instead of in separate rounds, ignore the first stitch when you reach the end of the first round, just work over the top of it and keep going. *Note.* Knitted spheres are dealt with in the last chapter.

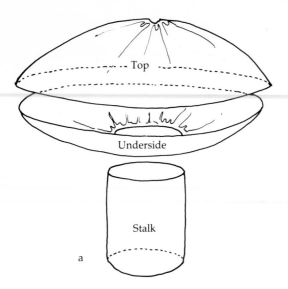

Top

Underside

Stalk

a

The three knitted or crocheted sections

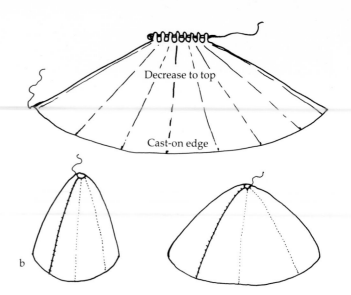

Decrease to top

Cast-on edge

b

Top sections made by decreasing

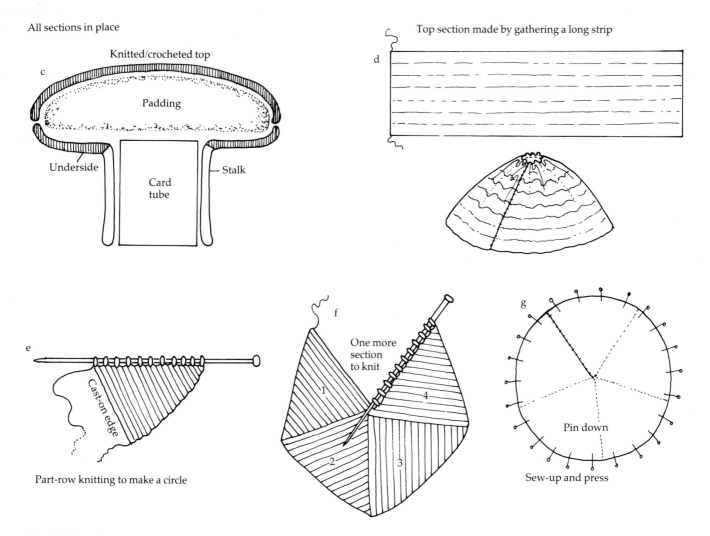

All sections in place

c

Knitted/crocheted top

Padding

Underside

Stalk

Card
tube

Top section made by gathering a long strip

d

e

Cast-on edge

Part-row knitting to make a circle

f

One more
section
to knit

1

2

3

4

g

Pin down

Sew-up and press

Fig. 73. Fungi.
a. The three knitted or crocheted sections.
b. Top sections made by decreasing.
c. All sections in place.
d. Top section made by gathering a long strip.
e. f. and g. Part-row knitting to make a circle.

44

Fig. 74. Knitted and crocheted fungi. The tallest one is only
3 in. (7.5 cm.) high.

FANTASTIC FUNGI

These knitted and crocheted mushrooms and toadstools are quite small – about as big as real ones – so
they only need small amounts of yarn and a few
stitches. The tops need no card stiffening, they will
usually keep their shape without that, but you may
wish to use a very small piece of terylene padding for
some of the domed ones.

Look in nature books for shapes of toadstools; there
are many varieties. Or you can make up your own,
using your favourite stitches, colours and yarns. To
prevent your toadstool from falling over, it can be
stuck on to a disc of green card.

The top and underside

These two circular pieces can be made in several ways.
Here are some suggestions:
1. Domes or discs of crochet in rounds (see page 43).
2. A long knitted strip, sewn into a tube and gathered
along one edge (d).
3. Cast on enough stitches for the outer edge, and
decrease regularly to the top. Sew up the side seams
(b).
4. Part-row knitting to make a full circle (e, f, and g).
5. Combine any of the above four methods, and add a
crochet frill around the outer edge.

Stalk

This may need a small tube of card to keep it rigid. If
you have one, you will want to make your stalk to fit
the tube, so measure its length and circumference
before you begin. Knit or crochet this:
1. In one long strip, joined to make a tube, in any
stitch.
2. In a tube as shown on page 43.

The patches of lichen on the log are explained on
page 46.

Part-row knitting

1. Cast on 10 sts.
2. * Knit to the last st., turn, and p. back.
3. Knit to the last 2 sts., turn, and p. back.
4. Knit to the last 3 sts., turn, and p. back.

Carry on in this way, leaving one extra stitch at the
end of each knit row, until there are no more stitches
to leave. * Your piece of knitting will then look like (e).
5. Knit across all the sts. and p. back.

Repeat the above instructions from * to * 4 *more* times
making 5 sections in all.
6. Cast off and sew up the two inside edges.
7. Pin the shape out to form a circle, and press gently.

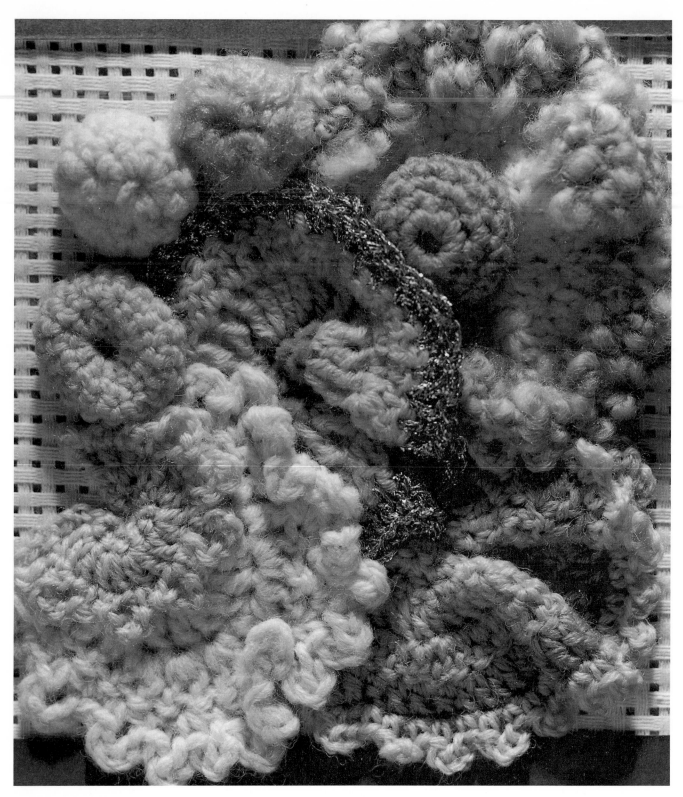

Fig. 75. Crochet domes, or cups, and ruffles.
Ruffles are made by working many trebles into the same
chain-space. Make a foundation chain of about 8, then work 3
or 4 tr. into each one, and continue to increase in this way on
every row. The patches of lichen on the log are made in this
way. (Instructions for the cups and domes are on page 43.)

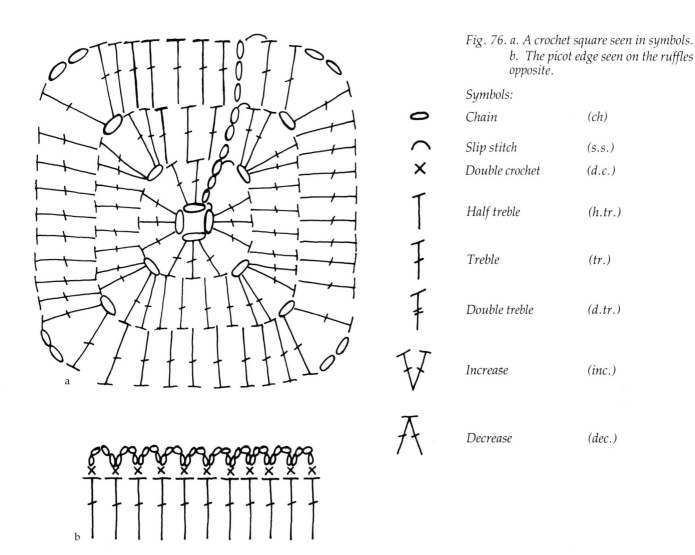

Fig. 76. a. A crochet square seen in symbols.
b. The picot edge seen on the ruffles opposite.

Symbols:

o	Chain	(ch)
∩	Slip stitch	(s.s.)
×	Double crochet	(d.c.)
T	Half treble	(h.tr.)
Ŧ	Treble	(tr.)
Ŧ	Double treble	(d.tr.)
Ⅴ	Increase	(inc.)
Λ	Decrease	(dec.)

CROCHET IN THE SQUARE

There are many different patterns for squares, but all are based on the same theme, that is, a foundation chain made into a circle, and an increasing number of stitches on each round. The difference with the square is that the increases all come at the same four places, and these become the corners. (If you increased at the same *six* places, you would get a hexagon!)

Simple symbols

It is often easier to look at a diagram of stitches than to read a pattern which many people find confusing. Each type of stitch has its own symbol (you will have noticed them at the beginning of this chapter) and every pattern can be drawn in symbols.

Look at the diagram of the square (Fig. 76) and begin 'reading' it from the centre. You will see the symbol chart alongside it. Work around it in an anti-clockwise direction, as this is the way the stitches are made. Follow these directions as you read it:
1. 4 chains, slip stitch these into a circle.
2. First round: 3 ch. to make the first st., 2 tr., 1 ch., *3 tr., 1 ch.* 3 times. S.s. to the first st. of that round.
3. Second round: begin with 3 ch., 3 tr. into next

3 ch.sp. Into each corner, work 2 tr., 1 ch., 2 tr. (i.e. right into the hole). Along each side, work 4 tr., (one into each ch.sp.). Slip the last st. of the last corner to the 3 ch. at the beginning.
4. Third round: begin with 3 ch., then work 6 trs. along that side. Then in each corner space, work 1 tr., 2 ch., 1 tr.

Along each side, work 8 trs. until you get back to the first stitch again, when you need to put in one extra tr. on that side before s.s. to connect.

This will produce a fairly plain square; if you wish to make it more interesting, you can use a different colour for each round. It does not matter where you begin each new-colour round, as long as you retain the correct number of stitches, but the best place is in the middle of one side.

Ruffles

The smaller diagram below the square gives you the instructions for the picot edgings seen on the ruffles in the photograph opposite. It tells you that into each tr. you should work a d.c. and 4 chs.

Fig. 77. Crispin the crochet cobra.

Fig. 79. Humbug!

CRISPIN, THE CROCHET COBRA

Crispin is a very friendly snake made from a tube of crochet in brightly-coloured D.K. yarns. His hood is an oval-shape of knitted Garter stitch, with button eyes and a tongue of ribbon. Make a tube basket for him in a neutral colour, perhaps with coloured bands, and coil your snake inside to await the music of his charmer. (Recorder practice?) A lid for the basket can be a crochet disc topped by a pompon. A smaller snake can be made from French knitting.

HUMBUG!

These humbugs are favourite English old-time sweets with a new look. Large ones make perfect cushions, smaller ones may be toys or fun-presents. Made up of two knitted or crocheted squares and sewn up in the special way shown in Fig. 78, they can be plain or patterned, real or fantastic. Seen from the top, they are:

1. Blue/green Humbug Award with a knitted-in anagram of the word 'humbug' designed to be hung around someone's neck!
2. Yellow/green crocheted textured stripes.
3. Red/pink crocheted squares.
4. Pink/yellow, eight small knitted pattern-squares.
5. Metallic yarn cable pattern.
6. Two diagonal squares in red, white and blue with gold braid.
7. Traditional black and white stripes; as always, two stuck together!

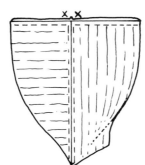

Fig. 78. Sew up three sides of the squares, pad the shape firmly, fold across in the other direction and sew up. Add pompon, tassels or braids as required.

48

Fig. 80. Cacti collection.

CACTI COLLECTION

The shapes of cacti are so easily and convincingly reproduced in knitting and crochet that, at first glance, the ones seen here may be taken for real. They are, in fact, little more than rectangles drawn up at the tops, or crochet domes, as on page 43. Their success also lies in the choice of yarns and stitches, many cacti having similar textures to furry brushed yarns and the regular pattern of knitted fabric.

These sit in the smallest size plant pots. Part way down from the top edge of each pot, a tiny ledge holds a piece of shaped card which has been stuck in place with glue. After padding, each cactus is stuck down firmly on to the card 'shelf', and where a gap exists around the edges, yarn looking like grit or soil has been added.

Notes on the 'Cacti collection'. From right to left, back row:

1. D.K. yarn, 3½ mm needles, 32 sts, double rib for 2½-3 in. (7 cm), k 2 tog. all along last row; sew into a tube and gather top and bottom edges. Crochet flower for top; beads sewn down each ridge.

2. 'Old Man' cactus; D.K. furry yarn, 4.50 mm hook, crochet rounds on about 20 sts as for domes on page 43. Brush with a wire brush to raise the pile.

3. Furry pale-green mohair with a slight slub, 4 mm needles, 40 sts. Reverse s.s. for about 3 in. (7 cm). K 2 tog. across the last 2 rows. Gather the last sts, on to a thread and draw up. Sew up and embroider pink French knots on top.

4. Two furry yarns knitted together, one plain, one random-dyed. About 30 sts on 4 mm needles, alternating s.s. and rev. s.s. Make a square piece and fold across so that the ridges are vertical. The red flower is just a bundle of red yarns, sewn on.

5. Front row left: crochet bobbles worked in rounds, using D.K. yarn, size 4 mm hook on about 18 sts, decreasing towards the top. The soft prickles of sewing cotton were attached after the padding stage.

6. Rectangle of single moss st., no shaping required. Slight padding in the largest piece, smaller pieces simply folded over. Yellow flowers are crochet cups with a little glitter yarn inside.

49

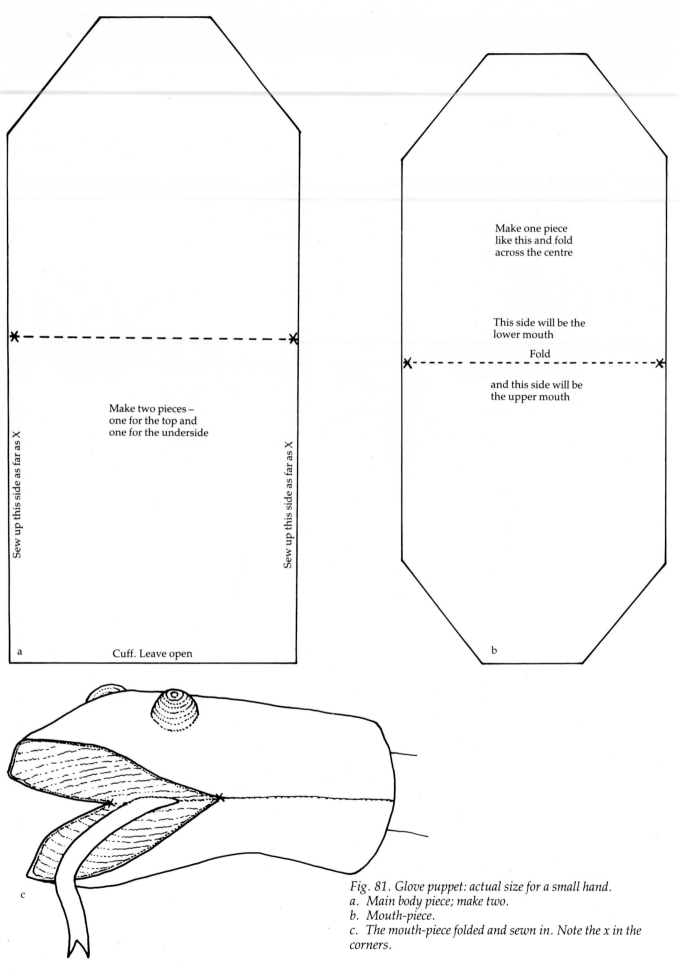

Make one piece
like this and fold
across the centre

This side will be the
lower mouth

Fold

and this side will be
the upper mouth

Make two pieces –
one for the top and
one for the underside

Sew up this side as far as X

Sew up this side as far as X

a

Cuff. Leave open

b

c

Fig. 81. Glove puppet: actual size for a small hand.
a. Main body piece; make two.
b. Mouth-piece.
c. The mouth-piece folded and sewn in. Note the x in the
corners.

50

Fig. 82. An assortment of simple glove puppets, some with mouths and some without.

GRUESOME GLOVE PUPPETS

Slip a hand into one of these appealing creatures and it will instantly provoke a response from both wearer and audience. The simple 'mitten' without the extra mouth can be any kind of creature, a bug, caterpillar, snake or worm. Add ears and a trunk and it could become an elephant, or two long ears and it will become a rabbit. Choose yarns which will best express the colour and texture of an animal, or let the yarn itself decide what develops. Pinky-beige for a pig with a ring at the end of his nose, marmalade (random-dyed) yarn for a tabby-cat, fluffy black and white for a dog with floppy ears. Dragons are universally popular among boys; these may have fearsome teeth and extra pieces sewn all over, bells and enormous eyes made from crochet domes. They can be benevolent or quite wicked – the latter may even need a knitted bag as his lair, perhaps with his name on to avoid mistakes! Naturally, this would be known as a Dragonbag.

The two pattern-pieces need little explanation. Knit two main pieces for each creature, remembering that, for some creatures, the underbody may be of a different colour and texture from the topside. Then, either sew these two pieces together all round, or make an extra mouth-piece and sew this along the mouth-edges as shown in the drawing. This is usually red or pink, but in the case of dragons and other monsters, any colour goes!

Metallic and glitter yarns are particularly good for this project, as they will glitter as the puppet moves. Fish, especially, look best when made with shiny yarns. Fins, gills and other pieces may be added for extra effect, and a row of fringing would finish off the sleeve edge with a flourish.

Eyes can be made from buttons, pieces of felt, sequins or beads, or slightly padded crochet domes. Tongues can be made of knitting, crochet, ribbon, braid or hand-made cord. Teeth (soft ones, of course) can be of shaped felt, though small beads and bells are also useful for this.

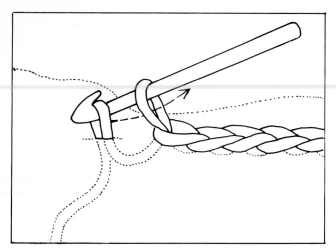

Fig. 83a. Surface chaining, first method.

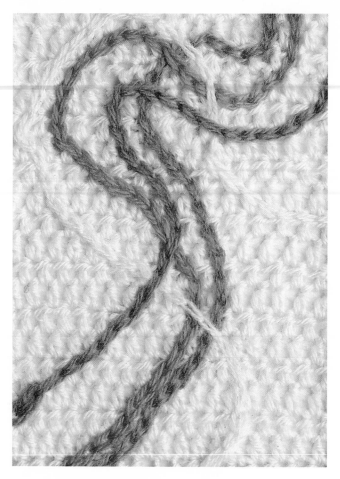

Fig. 83, above, and *Fig. 83a*, left. *Surface chaining worked on a background of half-trebles. The text explains how this is done and the diagram illustrates the first method described.*

SURFACE CHAINING

This is a useful way of adding lines, as part of a design, to any surface through which a crochet hook will pass easily, usually on crochet or knitting. Crochet stitches do not lend themselves easily to designs requiring very precise, wavy lines, so working them afterwards on the surface (rather like an embroidered chain stitch) is a good solution to the problem. There are two ways of doing this; only one way is shown in the diagram below. The first method gives a flat line, close to the surface of the fabric, whereas the second method gives a raised line because of the double yarn-over.

1. Hold the supply yarn *underneath* the surface of the work, and hold the end of it with the left hand ready for the hook. Enter the hook into the fabric, and yarn over (i.e. under the surface). Then pull it up to the top.

2. Now enter the hook again, the space of a chain away from the last one. Yarn over, pull the loop up to the top and straight through the loop on the hook.

3. Continue making the last movement, yarning over (under the surface) and pulling up and through both loops in the direction you wish your line to take. Finish off in the usual way, darning ends in on the reverse side.

The second method has the supply yarn *on top*, and so for this the hook has to enter the surface of the fabric and come back out again a short space away before yarning over and pulling through the fabric. Now with two loops on the hook, the stitch is completed as for double-crochet, that is, yarn over again and pull through two loops.

CHAPTER FOUR

Patterns in stitches

Patterns in stitches – Knitted buildings – Cable patterns – Simple cable exercise – Woolly jumpers (not for humans!) – Crochet stitch variations – Wall-hanging samplers – Piecework patterns – Lace knitting – Chevrons in crochet and knitting

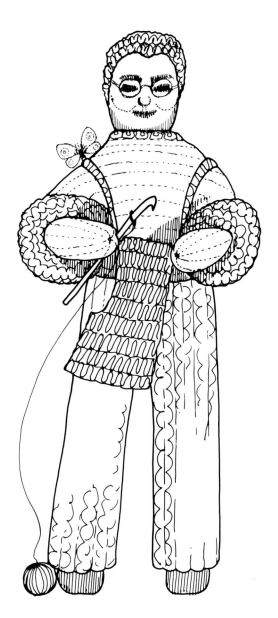

Keith is crocheting a waistcoat for his wife.

Patterns in stitches

All stitches, in both knitting and crochet, make patterns of some kind (even plain rows of stocking stitch) because each stitch-unit is repeated in a regular formation, side by side and one above the other. Introducing other elements into the stitches re-arranges them into a different pattern and we can easily see the changes produced by the effects of light and shadow on the surface of the fabric. These changes do not need colour to make them obvious, and so are called 'stitch patterns' as opposed to 'colour patterns'. Stitch patterns can be identified by touch alone, even without looking, whereas colour patterns must be *seen* to be appreciated.

Very simple changes to the knit stitch can produce effective patterns simply by adding purl stitches, increases and decreases, by putting the yarn in front or behind the work, by slipping stitches without knitting them, and by working stitches in a different order from the usual one.

In crochet, one can insert the hook into various places in the fabric, not only into the chain space below. This alone will produce a change of pattern, as does making taller and shorter stitches, bundling them together or spreading them out. These are all simple variations on the usual method.

Pattern is all around us: it is part of our everyday life, there for us to pick up and store in our memories, and to be used in our creative work. How much you notice and use it depends on your level of observation, and on your involvement with pattern-making activi-ties in other media, but if you knit and crochet you will no doubt have already noticed the strong similarity between the patterns you make and those seen on buildings. Pattern-writers must have noticed them too, for there are stitches named after parts of buildings, brick st., tile st., lattice and trellis, gothic windows, pyramids, pillar st., Eiffel Tower st., and no doubt many more in other countries. Indeed our choice of subject in decorative knitting and crochet may be partly governed by the close relationship between the patterns available to us in stitch manuals and those seen in our environment, on ice-cream cones, pave-ments and railings, in gardens, birds' feathers, woven fabrics, sculpture and many natural forms.

Scale is important too. When you knit or crochet a pattern you have seen somewhere outside, try to maintain the same scale as that seen on the original idea, so that, for instance, tiles and bricks are not gigantic out of proportion, or the feathers of a bird clumsy and untidy. Remember that if you want to reproduce a pattern fairly precisely, you should use a crisp yarn, and not-too-large tools, otherwise the effect becomes loose and the shape difficult to define. Of course this is not to say that all patterns must be worked on the same type of yarn. Experiment to discover other possibilities, but be prepared for some of your results to be better examples of *texture alone* than of pattern, and make a note of this in your notebook.

a

b

c

d

Fig. 85.
a. Stocking stitch, with reversed stocking stitch ridges. Alternate rows of knit and purl, with occasionally two knit rows and two purl rows together.
b. Right: Double rib; 2 knit and 2 purl alternately.
Left: Single rib: 1 knit and 1 purl alternately.
For these two rib patterns, work on an even number of sts, and remember that a knit stitch becomes a purl on the other side, and vice versa.
c. Top: Double Moss stitch: made like a double rib for two rows, and then the knit and purl stitches change places for the next two rows, and so on.
*   Bottom: Single Moss stitch: work on an odd number of stitches, and k1, p1 on every row, beginning each with k1.*
d. Waffle stitch: Cast on a multiple of 3 sts,
*Rows 1 and 3: *k2, p1* repeat.*
*Row 2: * k1, p2* repeat.*
Row 4: Knit. These 4 rows form the pattern.

Right. *Fig. 86. Buildings have many patterns on them which can be reproduced in knitting and crochet. Roofs, tiles, bricks, stones, fences and railings can all be used as starting-points for a design. Note the close relationship of these patterns to the stitches on the left.*

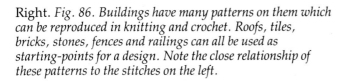

Above. *Fig. 88. Knitted buildings. A large panel of charming old-world cottages made up of simple rectangles of knitting, assembled and sewn onto a background of hessian (burlap). The tree and other foliage is crocheted, and also the smoke.*

Left. *Fig. 87. Knitted and crocheted building patterns, using mostly plain yarns with some random-dyed ones here and there. The roof is simple filet crochet with a decorative edging.*

KNITTED BUILDINGS

Many old buildings, particularly Victorian terraced houses, are very decorative indeed, making them very good subjects for knitting and crochet patterns. The house on page 56 has been constructed from only three main pieces, two long strips at the sides rather like samplers of stitches, and an oblong piece for the roof which owes its shape to the cut-out card which lies over it. The windows and doors are painted on the centre panel.

Very little shaping was needed to make the pieces for 'Knitted Buildings', even the doors and windows are made from oblongs sewn on to the main sections. Window and door-frames and all other details have been added, once the pieces were assembled on the background.

This would make an excellent subject for a group project in which even the youngest member could contribute a chimney – or even the smoke!

CABLE PATTERNS

Some of the most beautiful knitted stitch patterns are those produced by making two or three stitches change places with each other on the needle (i.e. crossing them over). To do this, we usually (though not always) use a cable needle, a very short needle with a point at both ends, just long enough to hold a few stitches at the front or back of the work while we knit the next ones. Then we put them back on again! Some cable patterns look very complicated, but once you know how it is done it is just a matter of practice before you become quite an expert.

The patterns along the top of page 59 show traditional designs which have been used by knitters in the British Isles for many years. Most of them are used by people who live near the sea, and they have been given names associated with the lives of fishermen and their families, like the simple cable pattern which resembles the ropes in daily use. The lobster claw, diamonds, moss, trellis and honeycomb are others. These old knitting patterns are very similar to the ancient strapwork and interlaced designs found on carvings all over the British Isles, and on some of the earliest manuscripts now in our museums. *The Book of Kells* and the *Lindisfarne Gospels* are two examples. These designs can be seen on ancient stone crosses, on church doorways and on fonts and pillars in the very oldest buildings in the country.

It may be coincidence that these patterns not only decorated stones and manuscripts in the distant past, but have been kept alive in knitting too. However, this will certainly ensure that they are not forgotten by future generations of knitters, and all those who enjoy wearing their own products.

In cable knitting, it is usual to use only one plain colour to accentuate the distinctive pattern, traditionally cream, neutral or navy. The best type of yarn to use is a firmly-spun, thick, smooth one called 'Aran' which gives a good definition to the design, though in knitting for purely decorative purposes you may like to experiment with others. However, if you are cabling for the first time, use a thick *smooth* yarn, as this is definitely easier.

Fig. 90. Stonework carvings showing Celtic designs on a church font and an Irish cross.

Page 59, opposite. *Fig. 92. A cable-covered tower made from an old hat-box. The brown arched doorway can be seen on the left, framed by two white cables. Panels of different cable-patterns are joined together all the way round; some of these joins are concealed by crochet edgings. The flat top is covered by a circular piece in the colours of lichen and moss, and cord threaded through holes in the top and bottom ties both sections together. The inside is lined with coloured knitting, the walls in a continuous strip of Garter stitch with Stocking stitch stripes. The lid and base were knitted on five kneedles though each one was started on a crochet circle from which stitches were picked up on to needles after the second round. This method is much easier than casting on to three or four needles to begin.*

Fig. 89. An ancient strapwork design from a Celtic cross, now in the Victoria and Albert Museum, London.

*Fig. 91. Traditional designs from fishermen's jerseys,
ganseys and Arans.*

*Fig. 93. The inside is lined with
coloured knitting.*

SIMPLE CABLE EXERCISE

As well as the usual pair of knitting needles, you will also need a short cable needle which should be either finer than, or the same size as the others you are using.

Cables are usually made in stocking stitch on a background of reversed stocking stitch, so the right side of the work (R.S.) will begin with purl stitches. The cables are always made on a R.S. row. The wrong-side rows are just knit and purl stitches, with no actual cabling involved.

Several rows are usually worked between each cable row: normally between four and six. The chart shown (Fig. 95) explains how to do a simple cable of four stitches (twice) with two stitches between each set. On the chart, the plain white squares represent knit stitches, and the shaded ones are purl stitches. Each square is one stitch (eighteen sts in a row) and this sample requires 24 rows. Begin at the bottom and work 2 rows. When you reach row 3, work as follows:

1. P4, slide the next 2 sts from the L.H. needle on to the cable needle and hold these at the *front* of the work. Knit the next 2 sts off the L.H. needle, then knit the 2 off the cable needle, and p2.

2. Now slide the next 2 sts from the L.H. needle on to the cable needle and hold this at the *back* of the work, knit the next 2 sts from the L.H. needle, and then the 2 from the cable needle, and p4.

3. This cable row is only worked three times in our sample, on rows 3, 9 and 15. The last 4 rows are in single rib.

You can use this as a pocket, or make two pieces and use them as a purse, or keep your sample in your experiments notebook, with instructions (Fig. 94).

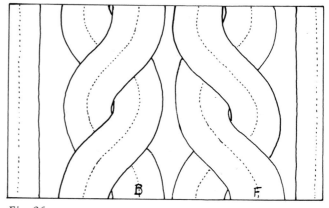

Fig. 95. Chart showing how to work a simple cable of four stitches with two stitches in between.

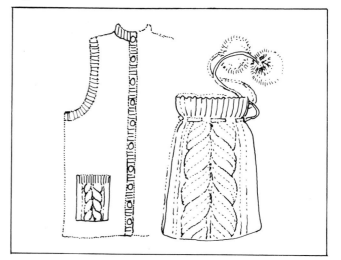

Fig. 96.

Fig. 94. Simple cables on a pocket.

Fig. 97.

Fig. 98. Woolly jumpers. A simple cable-pattern is used on the polo-necked jumper worn by the black-faced sheep. The large one with horns wears an 'Eyelet and Chevron' pattern,

and the one on the right is wearing a 'Vertical Openwork' pattern.

Below. Fig. 99. Diagram showing use of pipe-cleaners in the making of the sheep.

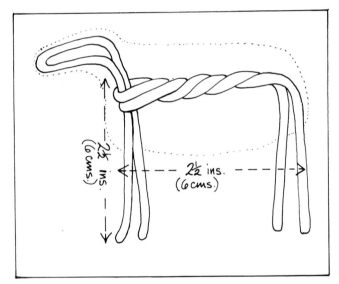

Fig. 99

Opposite page.

Fig. 96. A simple 2 × 2 cable pattern in diagram form, showing how the first two stitches cross in front of the second two, and in the next cable the first two stitches cross behind the second two.

Fig. 97. Two ideas for using the cable pocket, on a waistcoat and as a small bag. Make two pieces and sew them together along the sides and base, then make a cord and thread it through the ribbing at the top. The cord can be finished off with pompons or tassels.

WOOLLY JUMPERS (not for humans!)

You will need about 8–10 pipe-cleaners to give the bodies a framework for the padding and knitted jackets. Twist them together in pairs to aim for a shape like that shown in the diagram (Fig. 99), turning up the feet at the ends. The head and front legs are made of one set, and the body and back legs of another set. Extra wires can be wrapped around the body and shoulders. Pad the body thickly, as shown by the dotted lines, and wrap this in place with yarn. Wrap also the slightly padded head, and the unpadded legs. Make sure that it stands evenly at this stage.

Using fine wool and needles, knit the leg and face coverings from small oblongs, gathering the last stitches on to the nose and each leg before sewing up. The ears are made separately from about 4 sts.

Each sheep's jumper is made from a rectangular piece of knitting in cable and lacy patterns, and is slightly gathered along the centre back. The black one is still waiting for the shepherd to complete his coat, so he wears his own plain crochet rectangle until it is finished. The lacy patterns used are given below:

Eyelet and Chevron stitch

1. Cast on a multiple of 10 sts. plus one extra.
2. Knit rows 1, 2 and 3.
3. Row 4: K1,* y.fwd.,k3, k3tog.,k3,y.fwd., k1*.
4. Row 5: purl.
5. Row 6: as row 4.
6. Row 7: knit.
7. Repeat these 7 rows twice more, then continue only on rows 4 and 5.

Vertical openwork

1. Cast on a multiple of 4 sts.
2. Row 1: *k2, w.r.n., sl 1, k1, p.s.s.o. *
3. Row 2: *p2, w.r.n., p2tog. *
These two rows form the pattern.

CROCHET VARIATIONS

There are ways of making crochet stitches other than by putting the hook into the usual place (i.e. under the top two strands of the chain space). Instead, try putting the hook into the space at the side of the lower stitch, round the back and out at the other side (i.e. to the front) and thus make the stitch around the stem. This will produce Raised trebles (Fig. 100b.) The back of this stitch produces a pronounced ridge going in the opposite direction.

To make a Horizontal Ridge stitch (Fig. 100a.) put the hook into the *back* strand of the chain space only. This has a different effect on each side of the fabric.

Popcorn stitch (Fig 100c.) produces regular bumps by alternating tall stitches with short ones. This makes the tall stitches bend over into a small bump. If the tall stitches are placed exactly one above the other you will get an effect like that shown in the top half of the photograph. If the tall ones are placed *between* those on the row below, the effect will be the same as that seen in the lower half of the photograph.

a

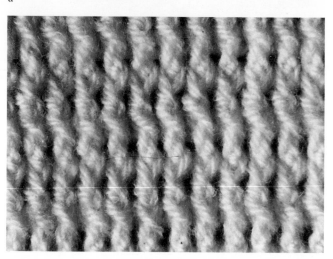

b

c

Fig. 100. a. Horizontal ridges
b. Raised trebles
c. Popcorn stitch
top: in ridges one above the other
bottom: the long and short ones staggered in each row.

WALL-HANGING SAMPLERS

Samplers like those shown on page 64 are an attractive way of practising stitches and keeping them handy for reference and decoration. Stitch patterns in knitting or crochet can be kept together in bands of different coloured yarns, rather like the knitters of Shetland kept their long samples of lace-knitting patterns for reference. The long samplers shown here have been folded into pleats and tunnels here and there, to take short lengths of wood and twigs, which is not only useful for keeping the sampler straight and flat, but also for disguising unsuccessful areas which you may prefer not to pull back.

You may want to make a sampler of cable or lace stitches to hang on your study wall as a decorative reminder of your newly-found skills. The ones shown here were made on about 20–24 sts, and can be as long or as short as you wish.

Piecework patterns

Here, below, are excellent ways of using small sample pieces of knitting and crochet to make cushion covers, bags, blankets, shawls, wall-hangings and panels. The two techniques may be mixed freely together using textured or plain yarns, in tones of one colour-family, or all white, black or grey.

The pieces in the drawing (Fig.101) are made up in the following ways:

1. Rectangles of mostly the same size.
2. Rectangles with one rounded edge, overlapping.
3. Rectangles with one pointed edge (like the bottom of the patchwork hanging on this page) (Fig.102).
4. Long strips, crochet squares, pompons and diagonal squares.
5. Crochet squares.
6. Assorted rectangles in knitting and crochet.
7. Four triangles, striped colours.
8. Four diagonal squares, striped colours.
9. Four small diagonal squares, four straight squares, and one crochet square.
10. One large square with a narrow strip of another colour in one half, used as a background for a line of four smaller sewn-on squares. The tassels are sewn along one line in the same colours as the squares.

Squares can be assembled to create colourful and attractive wall-hangings in a mixture of knitting and crochet (Fig. 102). This patchwork hanging is made from 45 × 3 in. squares (7.5 cm) in 4-ply yarns.

Fig. 102. Patchwork hanging.

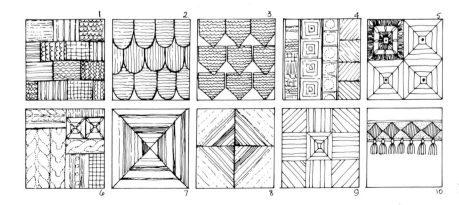

Left. *Fig. 101. Piecework patterns.*

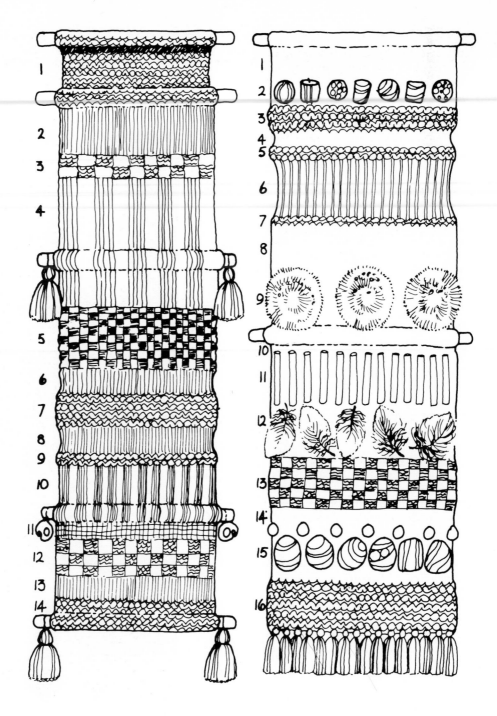

Key to stitches
Left sampler
1, 7, 9 and 14 Garter
stitch
2, 5, 6 and 13
Stocking stitch
3 and 12 Double
Moss stitch
4 Double rib
5 and 11 Single
Moss stitch
10 Single rib
Bells, tassels and
dowel

Right sampler
1, 4, 8, 10 and 14
Stocking stitch
2 and 15 Beads
3, 5, 7 and 16 Garter
stitch
6 Single rib
9 Pompons
11 Chopped
drinking straws
12 Feathers
13 Double Moss
stitch

Fig. 103. Sampler wall-hangings.

Fig. 104. One end of a barn near Henley-on-Thames, Oxon., which has been covered with old disused wall-paper printing blocks. This makes a most unusual wall-decoration of knitting and crochet-type patterns in a patchwork effect. (Photo by Valerie Campbell-Harding)

Fig. 105. A sampler of knitting stitch patterns in a patchwork formation exactly like that of the printing blocks above.

LACE KNITTING

All lace knitting is based on the various methods of increasing and decreasing the number of stitches on the needles, all of which are intended to create holes of different sizes. Many countries are well-known for their lace knitting, but especially famous are the islands of Shetland whose sheep produce an exceptionally fine wool most suitable for this kind of work. The traditional patterns used by the knitters of the Shetland Islands are worked on very fine needles (called 'wires') and very fine yarn; this is undyed and therefore always in its natural colours of creamy-white, grey, fawn, brown and black. Many knitters still use their own hand-spun for this.

The patterns are lacy and very beautiful, having such names as 'Old Shale' – a design which imitates the waves of the sea upon a shale beach. Other traditional names, which represent various aspects of nature in their design, include Ears o' grain, Cat's paw, Bird's eye, Fir-cone, Spout or Razor shell, Acre, and Horse-shoe. The patterns were not written down but were knitted in a long narrow sampler strip which was rolled up and kept for reference by the family. Each family would have its own slightly different way of interpreting the various patterns, and these would remain characteristic of that family.

Picot holes

1. One of the very simplest lace patterns for a complete beginner to try is a row of picot holes, which is very useful for threading ribbons and cords through on bags, and garments. They are made by alternately increasing and decreasing all along the row.
2. Cast on a multiple of 2 sts and work several plain rows before beginning the pattern.
3. On the right side * bring the yarn to the front of the work (this is called 'yarn forward' – y.fwd.) then k.2tog.* Repeat this to the end of the row.
4. On the wrong side, purl all the stitches – including the yarn-forwards, as these are actually put in to take the place of the decrease in the previous row. These are what make the holes. You should have the same number of stitches you began with.
5. The picot holes can also be folded across to make a decorative hem: the cast-on edge is sewn lightly to the wrong side.

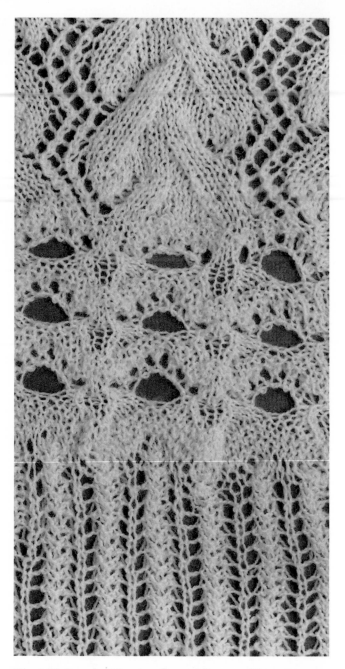

Fig. 107. Lace-knitting in a fine, white yarn. The patterns are named, from the top, Print o' the Wave, Crown of Glory and Razor Shell. Instructions for these can be found in many books dealing with traditional Shetland lace.

Fig. 106. Traditional lace-knitting patterns. Continued opposite.

Fig. 108. Corded rib.

Corded rib (Fig. 108)

1. Cast on a multiple of 4 sts plus 2 extra.
2. Work one foundation row of double rib, ending with k.2.
3. Every row after this is as follows:
k.1, * k.2tog.t.b.1. (i.e. through the back of the loop), wind the yarn round the needle to make one st., p.2, * k.1.

Note: it is only on the 2 knit sts where a decrease and an increase are made. To 'wind the yarn round the needle to make one stitch' you should bring it first *under* the needle to the front, then over the top of the needle, and back under to the front again ready to make the next purl stitches. This is abbreviated to 'y.r.n.'

Fig. 109. Garter Lace stitch.

Garter Lace stitch (Fig. 109)

1. Cast on a multiple of 2 sts.
2. Rows 1, 2, 3, 4, 5, and 6: knit.
3. Row 7. * Y.fwd., k.2tog.* repeat.
4. Row 8. * Y.r.n., p.2tog.* repeat.
5. Row 9. As row 7.
6. Row 10. As row 8.
These 10 rows form the pattern.

Note: to make a 'yarn round needle' at the *beginning* of a row when the first stitch is a purl stitch, enter the needle into the stitch purlwise (i.e. with the yarn in front) and hold both needles in position with the left hand while you take the yarn over the top of the needle, under it, and to the front again ready to make the next (purl) stitch.

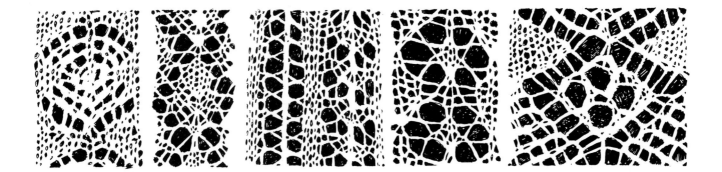

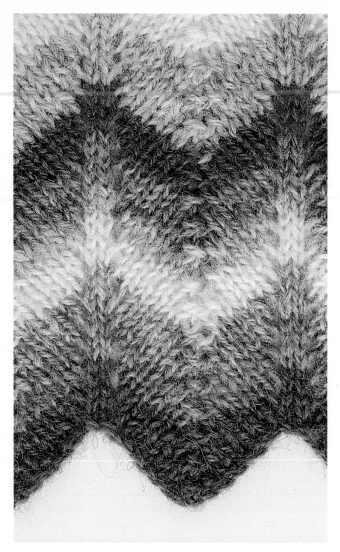

Fig. 110. Knitted chevrons.

Fig. 111. Crocheted chevrons.

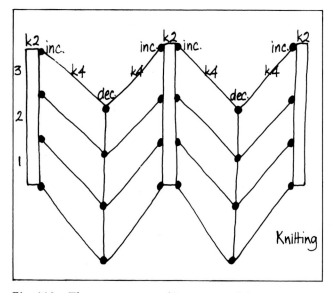

Fig. 110a. The arrangement of increases and decreases which form knitted chevrons.

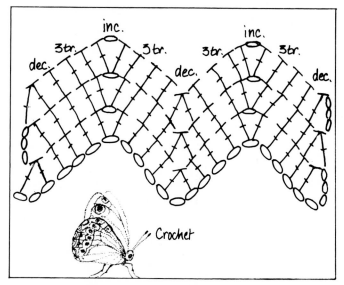

Fig. 111a. Crochet symbols which form chevrons.

Fig. 112. Garter chevron stitch.

Fig. 113. Fishtail stitch.

CHEVRONS IN CROCHET AND KNITTING

There are many variations on the chevron pattern in both knitting and crochet, some plain, some fancy, but all based on the same principle of increases and decreases in the same place on each row. Look at Figs. 110 and 111 and you will see that, at the top of each point, there has been an increase, and at the bottom of each point (the vee-shape) there has been a decrease. Always the same number of stitches is involved so that the number of stitches remains the same. Important points to remember are that:

a. more stitches than usual will be required when working chevron patterns, and

b. in knitting, the chevrons will not become obvious until several rows have been worked above them, as the needle holds the rows straight.

Knitted chevron pattern (Fig. 110)

Cast on a multiple of 13 sts plus 2 extra.

Row 1. * k.2, make one st.(m.1) from the loop between the next 2 sts, k.4, sl.1p-wise, k.2tog.,p.s.s.o., k.4, m.1 * to the last 2 sts, k.2.

Row 2. Purl.

These two rows form the pattern. In this version, the increases are worked on each side of the k.2 with which each knit row begins, making a vertical stripe of knit stitches between each chevron.

Crochet chevron pattern (Fig. 111)

The diagram of symbols below shows more clearly than words how to arrange stitches along the rows.

If you find it difficult to estimate how many chains you will need on the foundation row, the easy way round the problem is to make an entirely separate chain with its own yarn supply, and instead of beginning the first row as a continuation of the foundation chain, you start at the beginning of the chain leaving the end of it still attached to its ball of wool. As you go along, you will then be able to decide how many chevrons you need, and how many stitches, and you can shorten or lengthen the foundation row to fit by pulling some back or by adding some more.

Garter stitch chevron pattern (Fig. 112, page 69)

1. Cast on a multiple of 11 sts. Change colours whenever you wish.
2. Knit 5 rows.
3. Row 6. * k.2tog., k.2, inc.1 st. into each of the next 2 sts., k.3, sl. 1, k.1, p.s.s.o.* repeat to the end.
4. Row 7. Purl.
5. Repeat rows 6 and 7 twice more, then row 6 again. These 12 rows form the pattern.

Fishtail stitch (Fig. 113)

This very attractive lacy stitch is easy to knit.
1. Row 1. * Y.r.n. to make 1st., k.3, sl.1, k.2tog., p.s.s.o., k.3, y.fwd., k.1.*
2. Row 2 and alternate rows: purl.
3. Row 3. * k.1, y.fwd., k.2, sl.1, k.2tog., p.s.s.o., k.2, y.fwd., k.2.*
4. Row 5. * k.2, y.fwd., k.1, sl.1, k.2tog., p.s.s.o., k.1, y.fwd., k.3.*
5. Row 7. * k.3, y.fwd., sl.1, k.2tog., p.s.s.o., y.fwd., k.4.*
6. Row 8. Purl.
These 8 rows form the pattern.

CHAPTER FIVE

Creating with colour

Sylvia is knitting a lacy shawl from cobwebs.

Creating with colour

The choice of colours available to us in yarn shops is now greater than at any other time, so much so that we may sometimes be confused by shelves full of toning and contrasting schemes. Some of us may always use the same colour scheme through habit and a desire to 'play safe', and shy away from more adventurous combinations in case we get it wrong or feel too conspicuous or uncomfortable.

Nature never seems to get it wrong. Whatever colours appear together by accident, they never look out of place or discordant. If we try to analyse why this should be so, we would probably reach the conclusion that it has much to do with the proportions of each colour. The colours of nature which pop up together side by side are surrounded by a great expanse of another background colour, so that in context, they are only a very small proportion of what we actually see in one 'look'.

Left. *Fig. 115. Summer foliage.*

Fig. 116. Strands of yarns chosen to colour-match those seen in the photograph of summer foliage.

As an exercise, take a colourful photograph of plants and try to choose coloured yarns which exactly match them, in all tones and in the same proportions. Then try using all these yarns, again in the same proportions, in one of the projects, perhaps a humbug or one of the wrapped-yarn pictures.

In the picture of green foliage (Fig. 117) you will see a tiny patch of contrasting red. This has been reproduced in the knitted experiment (Fig. 118) as knitted strips of green with a tiny patch of red peeping through. This is a simple colour-exercise to use on something like a cushion cover or a bag. The strips have been woven together, and so several arrangements are possible.

Right. *Fig. 117. Autumn foliage.*

Fig. 118. A sample of knitted strips in the same colours as those seen in Fig. 117, woven to the same proportions.

EXPERIMENTS IN COLOUR-FUSION

There are various ways in which we can mix yarns of different colours and tones together without the use of a formal charted pattern. One is by alternating bands – which we call stripes – and another is shown on the previous page, in Fig. 118, where strips are woven together. In canvas embroidery and weaving we can easily change the stitch, or line, of colour at random, but in knitting and crochet we tend to feel that things should happen in a more organised manner because, traditionally, this is how it has always been done. There is no real reason why this should be so, as it is easy to create a random effect in knitting and crochet once you understand the tricks!

RANDOM-DYED YARNS

One particularly effective way is to use random-dyed yarns which allow you to make a colourful fabric without really trying. If you use one of these, together with a plain colour which is the same as one of the colours in the random-dye, you will find that part of your 'two-colour' pattern disappears where the two colours merge. Fig. 119 shows how the words appear and disappear where the grey patches in the random-dyed yarn coincides with the grey background.

Experiment in both knitting and crochet in order to discover how the patches of colour arrange themselves. In the knitted sample (Fig. 120a) a striped formation appears, and in the crochet sample (Fig. 120b) the colours form into blocks. This is because the stitches take up different amounts of yarn. You could try to *make* the colour patches stack up one above the other, or change stitches whenever the yarn changes colour or tone.

Use random-dyed yarns also as a way of moving gently from one colour into another, by choosing one which contains elements of both.

Other ways of fusing colours together softly are by knitting two strands of yarn together, changing only one of them at once. Using two strands of plain colour throughout this exercise, try the following order of colours in either crochet or knitting:
1. Two rows of white; cut one yarn and tie in pale grey.
2. Two rows white/pale grey. Cut the white and tie in the other pale grey.

Fig. 119. Patterns appear and disappear. Lettering, using random-dyed yarns on a plain background, showing how some colours coincide and disappear.

Fig. 120 a. Knitted stocking stitch showing how the colours in random-dyed yarns form lines.
b. Crochet sample showing how, in the same yarn, colours form blocks.

74

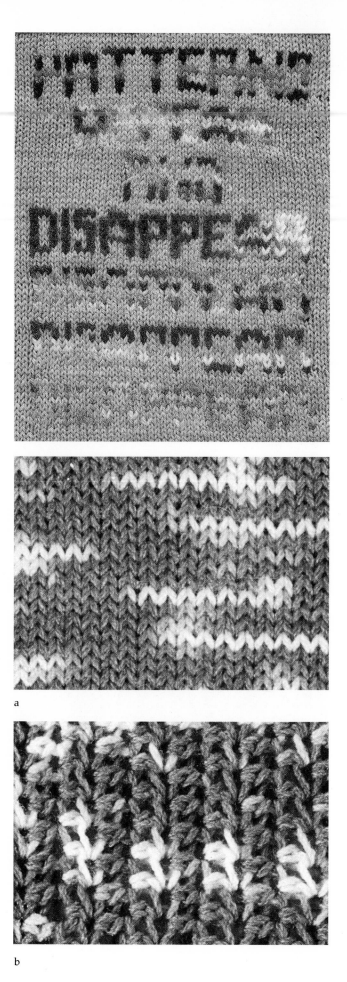

a

b

3. Two rows of pale grey; cut one and tie in dark grey.
4. Two rows pale/dark grey . . . Continue in this way, adding one strand of the next colour after every two rows, finishing with 'double black' for two rows. This should produce seven different bands, showing how four tones of one colour can produce nearly twice as many effects. If you used triple yarn instead of double, you would be able to produce even more tones.

Chequered stitches can produce a similar effect in a patterned way, especially in knitting where the units of pattern are smaller, allowing colours and tones to merge more easily. (See Fig. 121.)

In Fig. 122, right, you can see how, in crochet, the 'dropped stitch' can be used to overlap, or encroach, on to the row of a different colour below.

Begin by working two rows of trebles in the same colour. Then change colour and work alternately one d.c. and one d.tr. *BUT* make the d.tr. around the *stem* of the stitch which lies two rows below, that is, on the first treble row. Now work back on the next row using trebles in the same colour as the previous (dropped stitch) row, then change colours again. Now with the new colour (i.e. the third colour) work another row of dropped stitches, the long ones (d.tr.) going around the stem of the row two rows down, as before. This can be done by anchoring it around the short stitches or the previous long stitches. The example shows the dropped stitches going round the stem of the *short* ones in the relevant row.

Continue making alternate rows of trebles and dropped stitches, so that the encroaching stitches are only worked on one side.

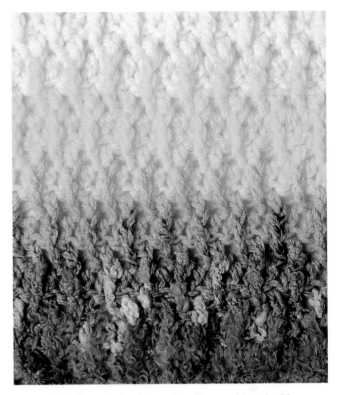

Fig. 122. Colour fusion in crochet. Encroaching double-trebles (d.tr.) made round the stem of alternate stitches two rows below.

This effect can also be similarly achieved in knitting, though here we have to take the colour upwards on to the rows above by holding stitches on the needle without knitting them. These are called 'slipped stitches'. Try a small experiment on about 12 sts. by working (in s.s.) four rows of one colour and then, on the next row, change to a different colour and slip one stitch in every three, except the last one. Work 2 more rows (i.e. 1 purl, and 1 knit) slipping the same sts. as before. On the next (purl) row, work across all the sts., and you will find that the slipped st. of the first colour has encroached onto the new one. Slipped sts. can only normally be carried for about three rows, as the fabric tends to contract and become denser than usual.

Fig. 121. Diagram of knitted chequered pattern showing how the colours can be arranged to fuse the tones.

POPPIES PANEL

The background of this colourful panel is one single piece of knitting using the colour-fusion methods of random-dyed yarns and the use of two strands together. Here and there, an extra strand has been woven across (using a needle) to introduce a paler or darker line, trying to avoid sudden changes of tone and hard unbroken lines of colour. On top of this background have been sewn frilly red and orange poppy petals, most of them crocheted, one or two in knitting. These were made on fairly large tools, in the manner of crochet 'rounds' with black-brown centres.

The secret of the colour-effect is to look very carefully at the *real thing*, and at many photographs, and to pick out *all* the colours you see there, in yarn. The temptation for beginners is to assume that 'poppies are red' and that any red is near enough. In fact, if we look more carefully, we can see that there are many reds, pinks, oranges and yellows in the scheme, and that the centres are not all black. The seed heads are tiny crocheted balls in fine crochet cottons and embroidery threads, stuffed with green wool, with chained stems. No leaves were made, as these would have covered up the interesting background.

The interesting frames were most important to the complete effect, and provide a link between the stone-coloured frame and the highly-coloured panel. The two wrapped arches are narrow shapes of card, cut with a Stanley knife. The same shape is cut from a large piece of card which covers the edges of the panel. The outer arched frame was wrapped with a speckled yarn of the same colour as the mount, to change the texture but not the colour. The inner arch is bound in the same colour as the panel, to relate in both colour and texture to the yarns on each side of it. These are stuck in place with strong glue. The bottom corners have been painted to match the colours of the yarns.

Notes on making the panel

Size 4.00 mm needles were used for the background, using many kinds of smooth and textured yarns of roughly D.K. thickness, on 64 stitches. Sometimes two thin yarns were used together. Only Garter stitch and Stocking stitch (right and reverse sides) were used for the background. There is no need to determine which is the right or wrong side too soon, but you may find that the 'wrong' side becomes the more interesting.

An extra 20 sts would have made this panel wide enough for the sleeve of a cardigan or jumper. It is impossible to estimate the amount of yarn you need when you are 'creating' as you go along, so if you are making sleeves more or less alike, work both simultaneously. However, they can look different!

Always leave enough extra knitting all the way round for mounting – about 3 in. (2 cm.). New threads introduced at the beginning of rows are left to hang loose until several rows have been worked. Then they are tied together tightly in twos. Ends hanging from the middle of a row can be darned in or tied later.

Bear in mind that rows of textured yarns, in normal lighting conditions, cast a slight shadow and so appear to be somewhat darker than they actually are. You can compensate by choosing one a tone or two paler.

Page 77. Fig. 123. 'Poppies' – a panel measuring 24 in. × 17 in. (60 cm × 42 cm).

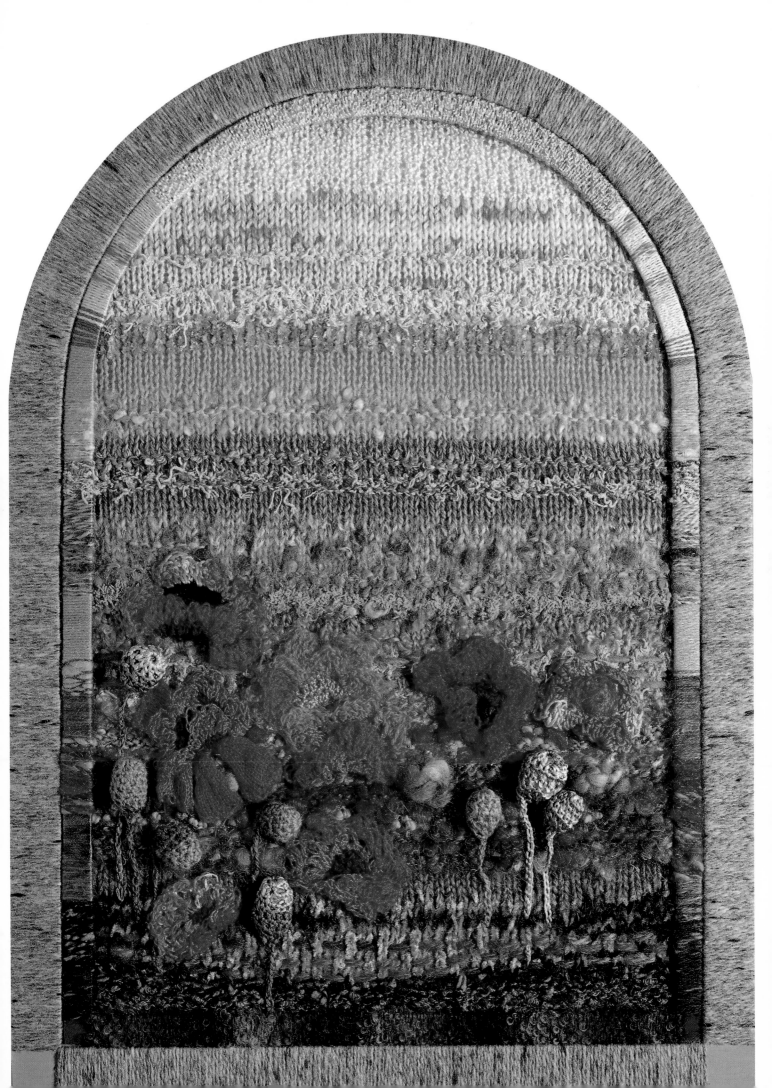

TWO-COLOUR KNITTING AND CROCHET

We have already used colour-changes in the previous projects by adding bands of different colours here and there; but if we use colours on the *same row*, simple patterns can be created. You can either use patterns which have already been worked out for you from traditional sources, or make your own on graph paper.

Here are some hints to remember if you are using two colours on the same row for the first time, or if you are teaching someone else to do this.

1. Work on very short rows to make sample pieces: about 20 sts. These can be kept, with notes, in your experiments book. If you wish to use them, you can edge them with crochet and sew them together to make a larger piece of fabric.

2. Use yarns which are of exactly the same thickness. If your work needs to be washed, the yarns should also be of the same type. Use needles of the correct size for the yarn, as the tension should not be too loose.

3. Work two or three rows of background colour before beginning the patterns.

4. In knitting, when you want to introduce the new colour, you simply 'drop' the old colour and pick up the new one to make the next stitch. In crochet, this is not quite the same. Fig. 127 shows how the last 'yarn over and pull through' of the previous stitch is made with the new colour, which then becomes the top of the new stitch. Try this to see how it works, and remember that, in crochet, every new colour is introduced 'early' in this way.

5. To make first attempts easier, use only shortish lengths of different colours instead of the whole ball, so that they will not tangle. These can be left hanging freely, or wound on to small pieces of card, with a notch to hold the yarn securely when it is not being used (Fig. 125).

6. Many traditional two-colour knitted patterns have only a few stitches between each change-over, so the yarn not in use is stranded across the back of the work loosely (see Figs. 124a and b). Take care not to pull these across too tightly, as this will make a difference to the width of your knitting. Five stitches is usually about the greatest number which should be stranded across without a danger of catching 'floating' yarns.

7. When designing your own patterns on graph paper for the first time, it is much easier to make every two rows the same (i.e. in pairs) which means that every purl row will then be exactly the same as the previous knit row. It will therefore be easier to follow without having to look at the chart. One simply has to follow the colours as they are on the left-hand needle.

8. There will be some distortion of every design as it is worked from the graph. This is because the graph chart is on squares, and knitted stitches are oblong! Therefore your knitted pattern will tend to look longer than it did on the chart, unless you compensate for this beforehand by making your design a little fatter.

9. It is possible to work motifs in two-colour crochet, but as the stitches are even more oblong than knitted ones, it takes a little more experience to plan a design. However, there is no such problem with border or all-over patterns.

Experiment to discover what is possible, and record your attempts.

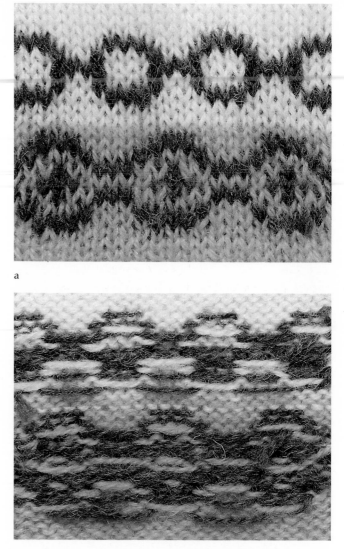

a

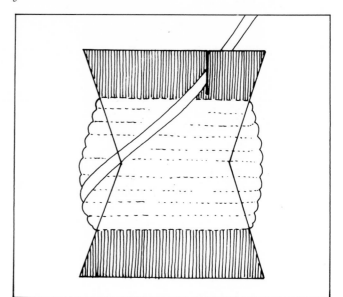

b

Fig. 124.a. A two-colour knitting pattern seen from the right side.
b. The same piece seen from the reverse side, showing the yarns stranded across the back.

Fig. 125. A small card yarn-holder, actual size.

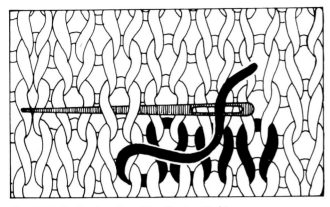

Fig. 126. Swiss darning (Duplicate stitch).

SWISS DARNING OR DUPLICATE STITCH
(Fig. 126)

This is an embroidery stitch which looks exactly like Stocking stitch. It is worked in a different colour on top of Stocking stitch to make shapes which are too tricky to knit in, for tiny motifs or letters, and other small details.

Use a large, blunt needle, and bring the point up through the centre of a stitch as shown, taking the needle behind the complete stitch above, and back down into the same place. Then move to the next stitch to be covered, and bring the point up through the centre of the stitch below and repeat. The new thread lies on top of the yarn which is already there, and the tension should be carefully adjusted after each stitch, to ensure that it lies flat on top.

The two motifs seen here may be embroidered on to knitted Stocking stitch using Swiss darning (Fig. 128).

Fig. 128. Charts for 'The Owl and the Pussycat'.

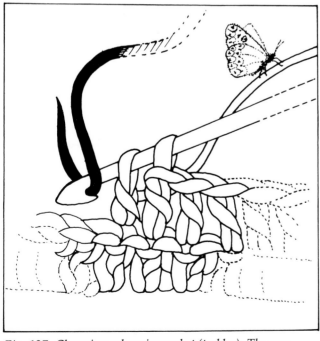

Fig. 127. Changing colour in crochet (trebles). The new colour is brought in as the last 'yarn over and pull through' of the previous stitch.

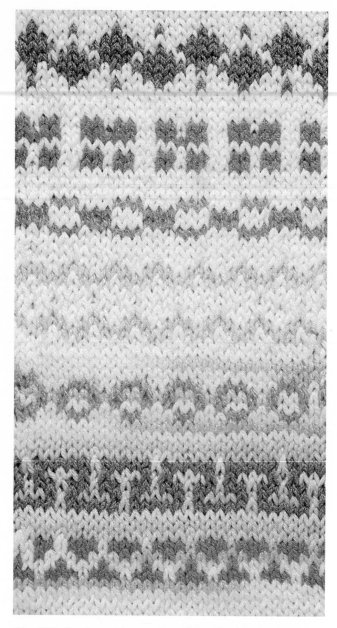

Fig. 129. Border patterns using plain colour and random-dyed yarns. The latter have been used sometimes as the background colour and sometimes as the 'pattern' colour. Only two yarns have been used for each border.

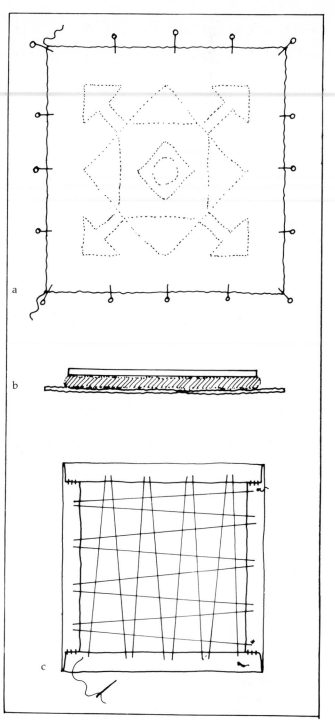

a

b

c

Fig. 130.a. Knitted square pinned out.
b. Padding sandwiched between knitting and card.
c. Knitting folded over padding and card, and laced across the back.

Fig. 131. The Fair Isle Printer's tray.

FAIR ISLE PRINTER'S TRAY

Many people will remember when printers used to keep their metal letters in shallow trays which were divided into compartments, some of them strengthened at the intersections with brass cross-pieces. These are now becoming obsolete in favour of computerised type-setting, and so the lovely trays can now be picked up in antique shops or markets, as dealers recognise their value as tiny shelf-units.

This one, bought at Covent Garden market, has been given a use in keeping with its traditional status. The small Fair Isle panels seem to reflect the geometrical nature of the box, while at the same time finding a way of displaying the lovely motifs as a sampler which will no doubt be useful.

Real Shetland wool was used for the project – spun in Lerwick, the capital of Shetland – and dyed in soft tones of brown, neutral, apricot and creamy white. The complete tray measures 15 in. × 13 in. (38 cm. × 32.5 cm.) and the larger pieces are only 3½ in. square (9 cm.).

As some of the squares are very small, it is best to use a fine yarn, such as a 3- or 4-ply, on size 12 or 2¾ mm needles. The long oblong section allows the chance to introduce border patterns, while some of the small sections are single motifs taken from borders.

The initials in the bottom left-hand corner can be worked out on graph paper once the number of stitches has been assessed.

The easiest way to begin is to knit one or two of the smallest pieces to assess your tension, and to gauge the size, allowing for about two rows extra top and bottom, and two stitches extra at each side. This is for turnings.

Each piece, after a gentle pressing is folded over a piece of padded card and laced across the back. It is important to cut the card pieces carefully, allowing enough room for the extra knitted fabric all the way round. The pieces are stuck lightly into each compartment with glue, though this is not essential, as they should fit snugly in position without dislodging.

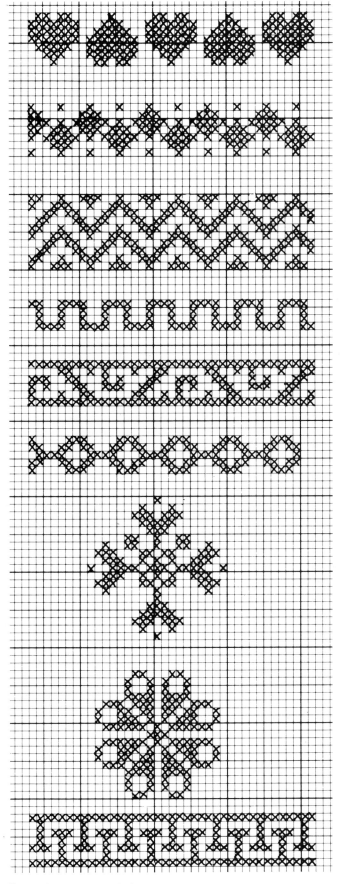

Fig. 134. The back of the Fair Isle box seen on page 30.

FAIR ISLE PATTERNS

The intricate and colourful patterns of Fair Isle knitting are world-famous for their diversity – there are about 160 of them, none of which is ever repeated on the same garment in exactly the same way. Using very fine needles (called 'wires') the Scottish knitters use Shetland yarn often in the natural colours of the fleece, white, black and moorit (brown) and with blends of these. In spite of the seemingly complicated patterns, only two colours are used on any one row, though the yarns are changed frequently to produce a rich, multi-coloured appearance. Designs have their special names, as in Aran and Shetland Lace knitting, and are known as Star of Bethlehem, Anchor, The Sacred Heart, and Rose of Sharon. Yarns are usually stranded across the back, as the intervals between colours are short.

Opposite page, and above.
Fig. 132 and 133. Patterns for two-colour knitting which can be used in a project like the printer's tray on page 81. These are traditional Fair Isle patterns. Each square represents one stitch.

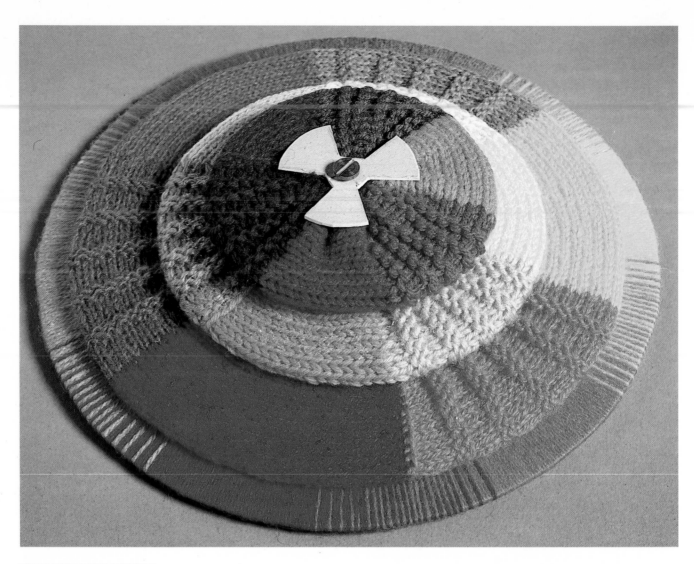

THE COLOUR WHEEL

This version of the colour wheel is not intended to illustrate the meaning and use of primary, secondary and tertiary colours (though it may also be used in this way as a bonus) but rather to illustrate the use of the technical terms such as 'tone', 'shade', 'hue' and 'tint'. These are words in everyday use which are often confused in application, though their technical meaning is quite specific. Making a colour wheel in knitted (or crocheted) yarns is a good way of introducing children to colour values leading to the understanding of colour schemes at a later stage.

HUE is the true basic colour: red, yellow, green, orange, violet, etc., without the addition of any other.

TINT is a hue to which white has been added to make it paler. SHADE is a hue with the addition of black (or any other dark colour) to make it darker.

All these are called TONES of a colour, and can range from the very palest through to the very darkest of the same colour, and there are many variations of these, depending on what, or how much, we add to them. For instance, we can lighten green by adding either white or yellow, and we can darken it by adding brown or blue or black.

Each of the discs on the colour-wheel represents the hues, tints and shades. For extra information, the primary colours are knitted in Stocking stitch, and the

secondary colours have bands of reversed Stocking stitch running across them. Primary colours – red, yellow and blue – are the only ones which cannot be made from combinations of any others. Secondary colours are those made by mixing together any two primary colours to make orange, violet and green.

The discs of card are covered by straight bands of knitting which are sewn into tubes, and slipped over the edges of each disc. The tubes of knitting are then gathered up to fit snugly as shown in the diagram (Fig. 136), each disc slightly overlapping the one beneath it to cover the gathered edge. A metal bolt will keep the discs in place, allowing them to revolve on the base. The base is a narrow ring of wrapped card stuck on to a disc of the same size. The coloured cover may be crocheted as well as knitted, and textured or random-dyed yarns may be used instead of plain ones. The number of stitches used for each band will depend on various factors: type of yarn, needles, tension size, etc. As a guide, the largest ring shown here (i.e. the hues) needed about 16 sts on size 3½ mm. needles in D.K. yarn. About 34 rows were needed for each colour section, but this should be measured from time to time against the lines drawn out on the discs.

If you prefer to make this a group project for children, each colour can be worked separately and then sewn together into one long strip.

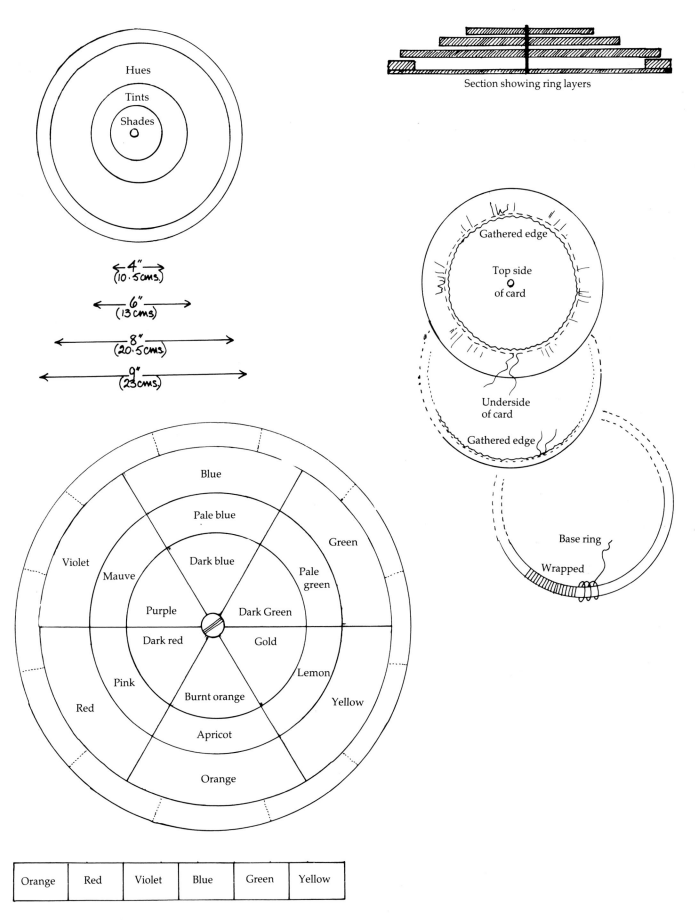

Section showing ring layers

Hues

Tints

Shades

4"
(10.5cms.)

6"
(13cms)

8"
(20.5cms)

9"
(23cms.)

Gathered edge

Top side
of card

Underside
of card

Gathered edge

Base ring

Wrapped

Blue

Pale blue

Green

Violet

Dark blue

Pale
green

Mauve

Purple

Dark Green

Dark red

Gold

Pink

Lemon

Red

Burnt orange

Yellow

Apricot

Orange

Orange	Red	Violet	Blue	Green	Yellow

*Fig. 135 and Fig. 136. The knitted colour wheel on page 84
illustrates technical terms like 'hue', 'tint', 'shade', and
'tone'. Above, a diagram of its constituent parts.*

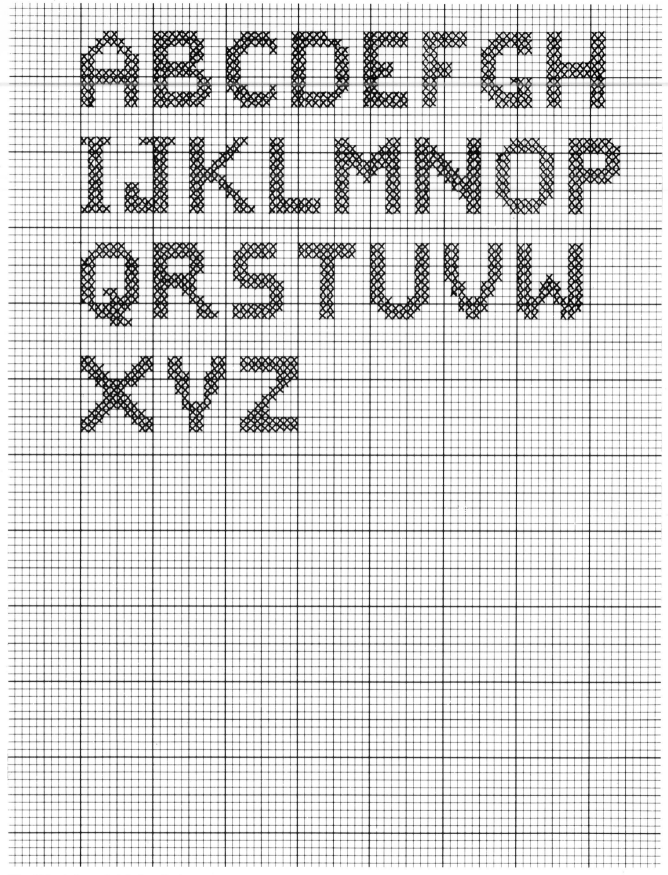

Fig. 137. A charted alphabet. Each square represents one stitch, and each letter is ten rows high.

CHAPTER SIX

Creating with texture

Ideas from nature – Creating textures with yarns – Textures in contrast – Knitted and crocheted Fur stitch – Knitted and crocheted bobbles – Moving from knitting to crochet – The 'House and Garden' panels

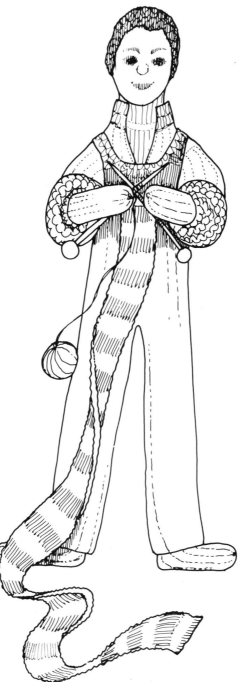

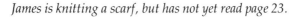

James is knitting a scarf, but has not yet read page 23.

Creating with texture

Fig. 139. Textured yarns, including metallic, bouclé, smooth-shiny, and mohair.

Fig. 140. Knitted string and leather, with hand-made beads. Different sized needles were used on this sample.

IDEAS FROM NATURE

It is a remarkable coincidence that many of nature's textures so closely resemble those made by knitting and crochet. Tiny lines and ridges, sometimes in formal pattern sometimes 'haywire', twists and bumps, furry patches – these can all be simulated in yarns using needles and hooks. In the photograph on page 90, you can see a detail of tree branches and trunk which have twisted together.

The sample of knitting on page 91 copies this by making long narrow strips with long slits in them – like extra-large vertical buttonholes – through which other narrow pieces have been threaded. Some of these are knitted in smooth stocking stitch, some in ridged Garter stitch, and some in textured yarns. It is useful to make samples like this, as ideas worked out in this way can be used on larger pieces. This happened to form ideas for 'Woodland Grove' (page 92) which is based on the same simple theme. It measures 17½ in. × 20 in. (44.5 cm × 51 cm.).

Creating texture with yarns

'Pattern' and 'texture' are two words which are closely related and yet which do not mean exactly the same thing. Pattern is made by units which are repeated, whereas texture has a surface interest which can be felt as well as seen. Sometimes the two words overlap, as they do in knitting and crochet, because knitted and crochet stitch-patterns can also be identified by touch. It is more difficult to identify the stitch when we use a textured yarn, because the extra sensations they produce in our fingertips confuse our knowledge of what we know of the stitch. As textured yarns produce a certain amount of surface interest on a piece of knitting or crochet, there is very little point of going to the trouble to use a complicated stitch pattern when the effect would be hidden by the yarn.

Materials

Try using 'alternative yarns' such as:
1. Narrow strips of cut nylon tights. These dye very well.
2. Strips of polythene cut from coloured carrier bags.
3. All kinds of string, dyed and natural. Soft rope.
4. Unspun or softly-twisted fleece.
5. Leather thongs.
6. Raffia and straw.
7. Ribbon
8. Soft wire, rubber and plastic tubing.

The drawings opposite show how your texture experiments can be mounted individually and hung together. Instructions for mounting small pieces are on page 80. Sew them on to fabric (curtain fabric, hessian and felt are the most useful) and decorate with buttons, beads, ribbons and knitted or crocheted edgings. The lower right-hand portion of the hanging shows realistic crochet grass with tiny white beads here and there to represent daisies. Instructions for knitted and crocheted loop stitches are on the following pages.

The cushions can, if you wish, be a mixture of knitting, crochet and fabric. Make the individual textured fish, plants, and trees, and sew them on firmly, or make a cable-log like the one at the left-hand side. To do this in a random fashion you need no written instructions, simply cable all over the place on narrow strips of different widths and then sew them all up together lengthways to make one long tubular piece. The ends are crocheted rounds.

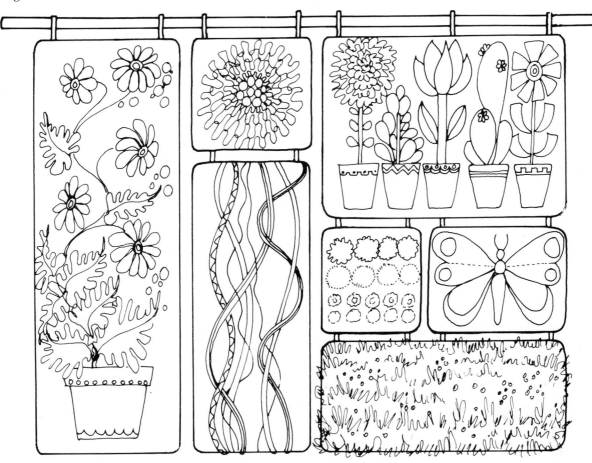

Fig. 141. Ideas for textured hangings and cushions.

89

Fig. 142. Tree bark.

Fig. 143. Strips of knitting twisted together.

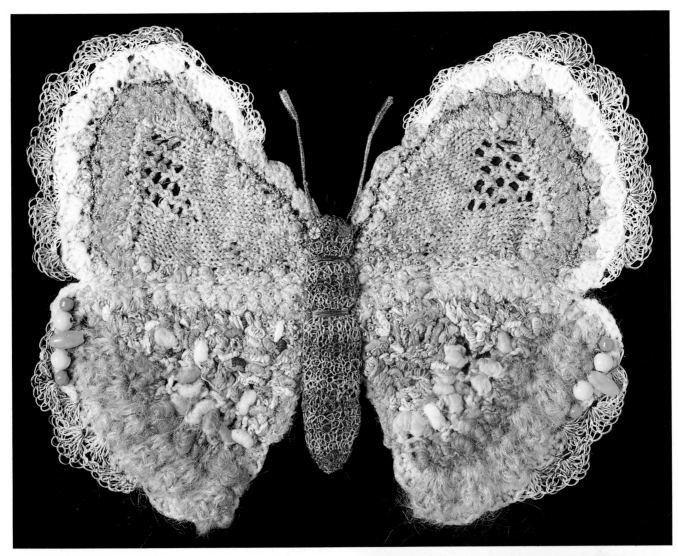

Fig. 145. 'Butterfly', made in a combination of knitting and crochet on a variety of different-sized needles and using many different kinds of yarn.

Left. Fig. 144. 'Woodland Grove'.
This panel is a mixture of knitting and crochet, with padded fabric. Most of these pieces are long strips twisted together and sewn on to a background of net over hessian (burlap).
Right. Fig. 144a. If one compares this photograph of tree bark with the knitting sample, Fig. 143, and with 'Woodland Grove' opposite, the logical progression from idea to sample to full-scale work is strikingly revealed.

TEXTURES IN CONTRAST

Everything we touch has a texture, smooth as well as rough, silky, hairy or bumpy: it is the degrees of texture which vary. We can, however, accentuate the texture of a yarn by the way we contrast it with another one. The full richness of thick bouclé (curly) yarns like the ones in the samples opposite on page 95, are more pronounced when placed side by side with smooth ones knitted in a simple stitch. In the top sample, even more contrast is supplied by the small rings which have been wrapped with different yarns, some shiny, some matt. Only white yarns have been used, as colours would have created confusion with so many different textures.

The lower sample is made of long narrow strips of Garter stitch in smooth and rough yarns. The vertical ones have been stapled on to a piece of card so that they form loops and tunnels for the horizontal strips to pass through. It is made in tones of violet and white.

When using a yarn which has occasional lumps of fibre along it (called 'slubs') you will find that these tend to fall on the reverse side of your work when you use a knit stitch or when you crochet. This is because your supply yarn is fed in from behind the work, and so the bumps get caught on that side. Sometimes these can be pushed through to the other side when only a few are involved. It is better to use reversed Stocking stitch for the right side to obtain the full effect of the textured yarn, and in crochet you can often manipulate the slubs as you go along, to make them fall on to the right side.

Very highly-textured yarns tend to hide the shape of the knitting or crochet stitch, and therefore disguise the technique which was used to make it. Therefore it is more sensible to use a simple stitch in whatever technique you find easiest. Whilst some stitch patterns will still be visible when made with a textured yarn, many will be quite lost, so, to avoid disappointment, make an experimental piece before you begin a large project.

Experiment to create your own textures in decorative projects by pleating, folding, smocking (on ribbed knitting), gathering and padding, just as embroiderers do on woven fabric. You can also make textures in smooth yarns by making bumps and bobbles (as in traditional Aran knitting and crochet) also fur stitches, pebbly patterns like Moss stitch, cross-over ridges like random cables (that is, cables done irregularly without the usual counting), and by making irregular holes, even by dropping stitches and by winding the yarn several times round the needle in knitting. Holes in crochet are even more easily made than in knitting, by missing chain spaces and making a bridge of chains followed by a tall stitch (e.g. a double-treble). Make many irregular-shaped ones next to each other to resemble the textured pattern of soap-bubbles.

For ideas, look at the texture of everything around you, and at photographs in books, especially those of nature. Try to imagine the textures seen in everyday life as pieces of knitting or crochet, and analyse them with stitches in mind. Use words like 'ridged, basket-weave, mossy, grassy, bumpy or furry' to describe what you see, and then make small experimental pieces to simulate these textures. You will find that, with the right combination of yarn, stitch and technique (i.e. knitting or crochet), the most exciting results will emerge.

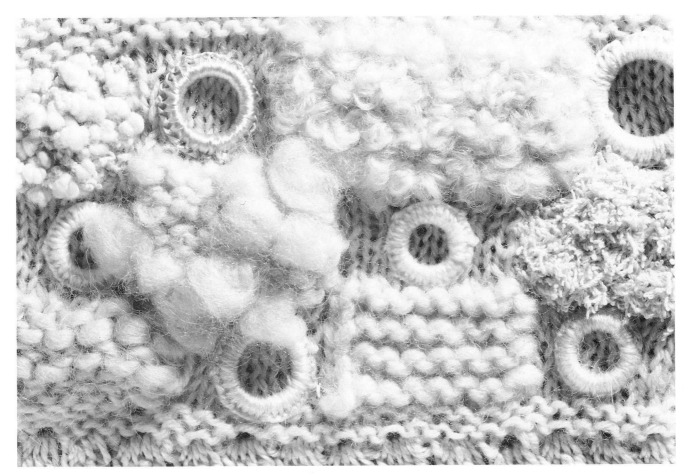

Fig. 146 and 147. See text opposite.

KNITTED AND CROCHETED FUR STITCH

Generally speaking, Fur (or loop) stitches, that is, those which can be cut to make fur effects, are easier to crochet than to knit. However, one way of knitting a Fur stitch is included here because it is a simple way round the problem for beginners and young people who like quick results. All Fur stitches are made from the reverse side as the loops fall on to the right side as work progresses. This means that every other row is a plain row used to lock the loops firmly into position, starting the loop row at the same end each time.

1. For the knitted version (Fig. 148) prepare a pile of short lengths of yarn, about 6 in. (15 cm.) long (your fingers will find it difficult to handle pieces shorter than this) and work in Garter stitch as follows:

1. Knit one stitch, *place one or two strands of cut yarn across the row between the needle points so that one end sticks out at the front and one at the back of the work.

2. Now knit the next stitch.

3. Pull the end nearest to you (i.e. of the loose strand) around the st. just knitted, between the needles and over to the back (i.e. the R.S.) so that the two ends now hang side by side.

4. Knit another st. and repeat from * to the end of the row. Knit back.

Note: while you are doing this, try to keep the supply yarn well out of the way so that it does not become entangled with the short pieces.

This is a very good stitch to use with short lengths of unspun fleece. Try knitting a beard, moustache or wig!

For the crochet Fur stitch, double yarn produces a thicker and denser fur than thin or single yarn. This is also worked from the back, and on *alternate rows* of double crochet stitch. This is how to make it:
With the back of the work facing you, enter the hook into the chain space as usual. Extend the yarn high up on the index finger as shown in the diagram. Move the hook around to the back of both strands of yarn and gather them in. Pull both strands through the chain space towards you. Now drop the loop and hold it down out of the way with your left hand. With 3 loops on your hook, yarn over, and pull through all three loops at once. Repeat this action into each chain space to the end of the row, then work a row of d.c. back again.

In both the knitted and crochet versions, the loops can be either left as loops or cut as short as you wish.

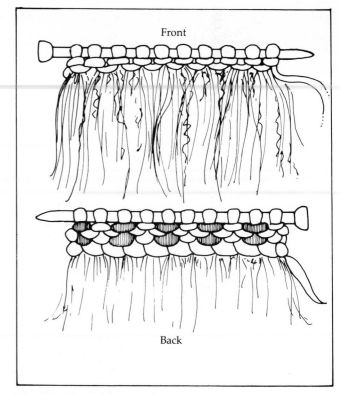

Fig. 148. Knitted Fur stitch.

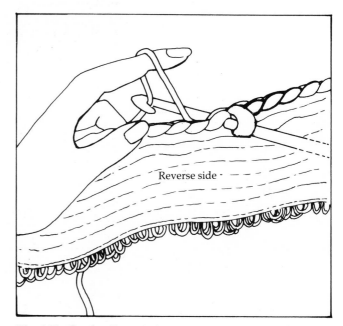

Fig. 149. Crochet Fur stitch.

Page 97. Fig. 150. 'Cats', 15½ in. × 75 in. (38 cm. × 187½ cm.).
A hanging of knitted and crocheted pieces applied to a background of hessian. The knitting and crochet is worked in all directions, picking up stitches from all sides and crocheting on to edges. The main bodies are one piece, but the faces, ears and eyes, etc., are small pieces made separately and then sewn into position. Note the different patterns made by the random-dyed yarn on the back of the centre cat, one side knitted and the other side crocheted. As with the other hangings, all kinds of yarn were used, thick and thin, natural and synthetic, and a wide range of tools and sizes.

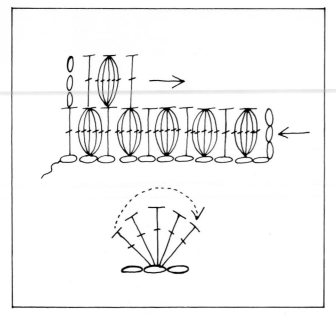

Fig. 151. Diagram in symbols, of bobbles in trebles.

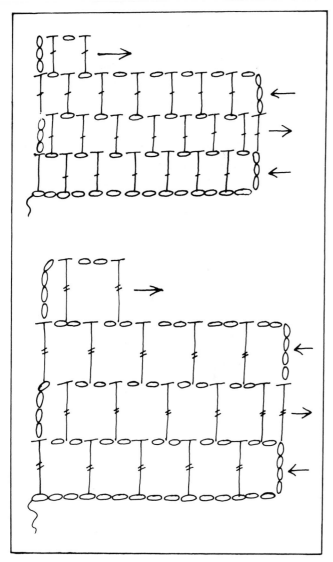

Fig. 152. Diagram in symbols, of filet crochet showing
a. trebles, missing one chain space;
b. double-trebles, missing two chain spaces.

CROCHETED AND KNITTED BOBBLES

Crochet

This is an extremely useful and attractive stitch, and essential for creating high textures. This crochet version is easier and quicker to make than the knitted version, which is included here for comparison.

For the best effect, use a firm, thick yarn, those with slubs and variegated colours are also good as they will produce slightly different sized and coloured bumps. You can make bobbles with double yarns too for even more variety, and try organising them in irregular clumps along the row, instead of evenly.

The diagrams show the symbols for chains and trebles, as these are easier than words to understand at a glance. The stitch is made by working 4 or 5 trebles into one chain space to make a fan shape – which is then closed at the top (see the lower part of the diagram). To do this, pull the loop up very slightly from the *last stitch* of the fan-shape, and remove the hook. Now insert the hook under both threads of the first stitch of the fan-shape from front to back, then back into the loop of the last stitch (which you pulled up). Now tighten the yarn slightly, and pull the loop of the last stitch through the two threads on the hook. You have now pulled the last stitch through the first one, closed up the fan-shape and made a bump. To make one on the other side of the fabric, insert the hook into the first stitch of the fan-shape from back to front.

Filet crochet

You will have noticed that when you miss a chain space and go on to the next one, you make a hole. This – making holes – is the basis of filet crochet. Of course, if you just continue missing chain spaces you will also be decreasing, so in order to keep the same number of stitches you must make a chain over the space that you have missed. If you miss more than one chain space, you must then make the same number of chains over the top, like a bridge. If you use tall stitches in between (double-trebles, for instance) you will create quite large holes. As an experiment, try doing this at random, some large holes and then some small ones next to them. Make the stitches of the next row *either* into the holes of the row below, or into the chain space.

Note that, in the lower diagram of larger holes, the double-trebles are worked into the complete hole below, not into a chain space, for, with only two, there is no centre chain.

Knitted bobbles

The small version is called Trinity stitch, and is much used in Aran designs. Its name derives from the method of making 'one into three, and three into one'. *On alternate rows* the method is to make three stitches in one (i.e. k.1, p.1, k.1, all into the same stitch) and then purl 3 together. This is done all across the row to the end, thus alternately increasing and decreasing on a multiple of four stitches. Make the bobbles of the pattern row fall between the previous ones by beginning with the decrease the second time.

Fig. 153. A detail from a larger panel of knitted and crocheted trees. Many kinds of thread have been used to vary the textures, including wool and fine crochet cotton on a background of striped furnishing fabric. You will be able to see quite clearly where the knitting and crochet take over from each other, and how filet crochet has been used in a loose way to create pattern and texture.

Larger knitted bobbles are made by increasing five stitches into one, knitting four rows *on these five sts alone*, and then passing them over the top of each other to decrease back to one again. This is done all along the row, on every fourth row. Unless you can learn to knit four or five stitches backwards fairly quickly, there is much turning of the work to make this stitch, and this makes progress rather slow.

MOVING FROM KNITTING TO CROCHET

It is easy to see how the two techniques can be used together in a decorative way, one to create a smooth, close texture, and the other to make loose holes. In the photograph above, you can tell which parts are knitted and which are crocheted. How to change from one to the other is explained on page 42.

Casting off knitting with a crochet hook is a good way of making the edge more suitable for crocheting on to it, as the chain spaces are constructed in a different way from a knitted cast-off edge. Simply take the knitting in the left hand as usual, with the supply yarn also in the left hand, as in crochet. Now, with a crochet hook, slip two stitches off the needle on to the hook, yarn over and pull through. Slip another stitch off the needle on to the hook, yarn over, and pull through. Do this to the end of the row; then, instead of cutting off the yarn, you will find that you are now crocheting instead of knitting!

Note: if you follow these instructions for casting off, you will have a looser cast-off edge than normal. To avoid this, you may prefer to use a smaller hook than usual, or slide off two stitches together out of every four, and cast them off at the same time. This will reduce every four stitches to three.

Fig. 154. The four 'House and Garden' panels, which are desribed on page 102.

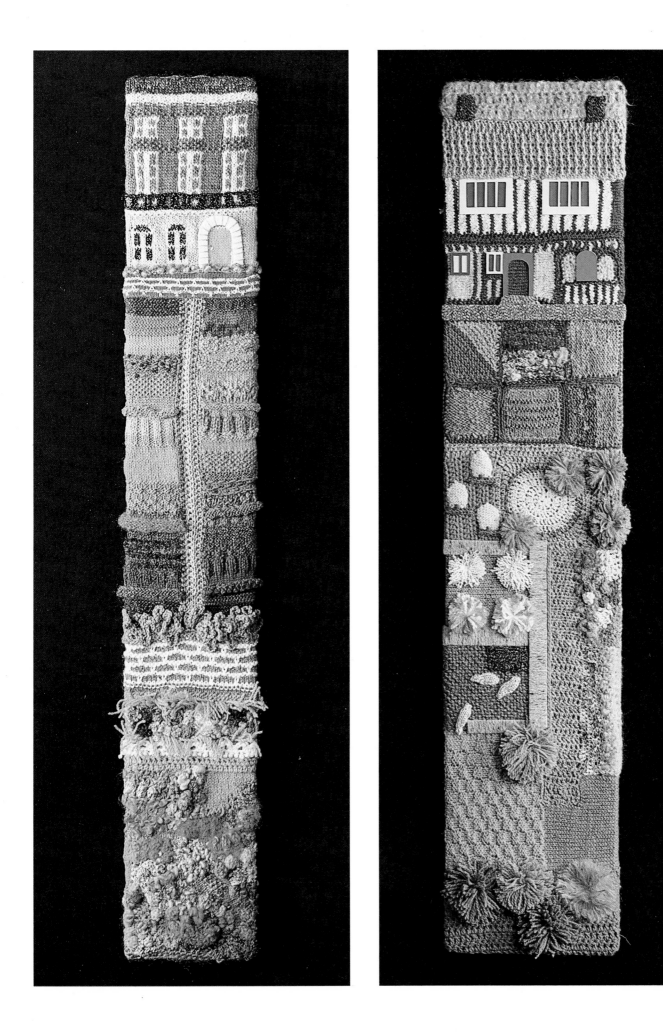

101

THE 'HOUSE AND GARDEN' PANELS (Fig. 154, pages 100–101)

Four long narrow panels made for the Knitting Craft Group stand at the Needlecrafts '85 Exhibition in Brighton. The widest ones are 15 in. (38.5 cm.) wide, and the narrowest 11 in. (28 cm.). They were made by the author, with additions from Sylvia Cosh and James Walters. Each panel is made of thickly-packed wood chips which provides a sturdy and firm surface suitable for taking staples. In each case, the background pieces (grass, paths, gardens, etc.,) were made separately, and either sewn together or stapled directly on to the board mounts. The long narrow format is one often seen by rail passengers who can see the backs of houses and gardens of this shape, row upon row, side by side, neatly packed together. An idea like this would make an excellent class project, as even the tiniest addition would make a valuable contribution to the whole effect.

Panel 1

Tiny vegetables and seedlings, crocheted directly on to the brown soil, or made separately and sewn on. Crocheted path and borders of blue and white yarns. Pale striped lawns of crochet, above, with lots of pompon bushes in mixed colours.

Panel 2

From the top: canvas-embroidered sky, cottages and stream, crochet Fur stitch grass above and below the water. Crochet tree and shrubs, pathway and greenhouse. Pompon foliage at base. This panel fits (quite by chance) exactly on to the top end of Panel 1, which becomes an extension of the garden.

Panel 3

A knitted town house, with added crochet balcony. The windows were Swiss-darned and the door added later. The walls are knitted, as are the two green strips at each side of the crocheted pathway. The foliage here is crocheted at random (a good place to practise crochet stitches, as mistakes do not matter – and indeed add to the textural effect!) and crochet ruffles form the foliage on top of the lower wall.

Panel 4

The sky and roof are crocheted, but the half-timbered effect on the house is knitted-in, except for one or two of the doors which were added later. The upper windows are painted card, and the path across the lower edge of the house is of wrapped card. The herb garden is made of knitted squares, below which is the crocheted pathway and pond. Beehives are made separately, as are the pompon bushes and fruit trees, and knitted pigs are about to escape on to the knitted common.

CHAPTER SEVEN

Creating in three dimensions

Tasty vegetables and fruit – Free-standing structures – Knitted
spheres and other ideas – Hanging structures – Costume
gallery – Little people – Large people – Working notes

Robin the Gnome is knitting his other boot.

Fig. 156. Tasty vegetables and fruit. The textures of certain fruit and vegetables are very suitable for rendering in crochet or knitting as indeed are their interesting variegated shapes. See page 106.

Tasty vegetables and fruit

The textures of some vegetables and fruit are perfect for translation into knitting and crochet, as are most of the simple shapes – round, long, pointed or bulbous. The ones on pages 104–5 are particularly simple, some being based on the ball shape with a few additions such as outside leaves. Those not shown could include tomatoes, beetroot, oranges, grapefruit, apples, potatoes, pomegranates and plums.

The carrots, like the leeks, are rectangles with shaped ends and gathered tops. The leeks have a small disc of card at the flat root end to keep them in shape. Notice how the tones of green have been graded, as explained on pages 74–5. The onion was started at the base, and stitches were increased quickly to the maximum number around the outer edge. These were then decreased towards the top. Yarns dyed in onion-skins were used for this. For the crocheted centre of the cauliflower, textured yarns were used and, like the lettuce, this has extra leaves sewn on to the base and sides. The instructions will explain how these were made. If you wish, add some natural-looking mushrooms, explained on page 44–45.

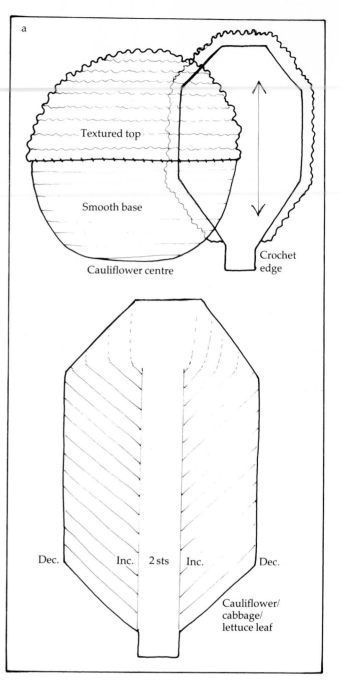

Cauliflower

The centre: make two domes (see page 43), one in a white textured yarn for the top half and one in a smooth yarn for the base. Fill these with Terylene padding and sew them together (Fig. 157a).

To make the individual leaves, use various tones of green, in a not-too-thick type of yarn. The shape is shown in the diagram, beginning with the stalk and increasing on each side of the central two stitches, while at the same time decreasing on the outer edge to keep the number of stitches the same. This gives a centre ridge to the leaf and makes the stitches follow the natural direction of the veins. The size of each leaf will depend on the size of the central section; the relative proportions are shown in the drawing. The frilly edge is made by crocheting two trebles into each chain space all round the edge, except for the stalk. Use a slightly darker yarn for the outer edge, and textured too. Five or six leaves will be needed, each one slightly over-lapping the ones on each side. These should be pinned in place and stitched on with yarn of the same colour.

Lettuce

The construction is the same as for the cauliflower except that the central ball is smooth and so can be made in one piece from the same yarn. Do not forget to pad it firmly before you close it up. The inside leaves, which you can see nearest the heart of the lettuce, are made from pieces like those in the small diagram. The long piece is folded across and stitched around the ball with the edges over-lapping, and the folded edge

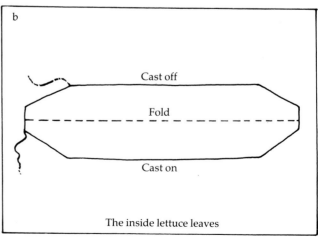

Fig. 157.

nearest the top. The outer leaves are the same as those for the cauliflower except that a brighter green should be used.

Leeks (see Fig. 158)

These are knitted in Stocking stitch, and you will need several tones of green yarn ranging from white to dark green. Where the tones change, use either double yarn, or the chequered effect as explained on page 75 to avoid a hard line of colour. Begin with the white yarn and about 25 sts in single rib; and continue with bands of deeper greens towards the top. To make splits in the top, only a few sts are worked at a time, casting off, and rejoining the yarn again for each new section until all the sts have been worked and each part is more or less the same length. Sew up the side seam, taking care to match the bands of green; the top 2 in. (5 cm.) can be left unsewn to make another split. The leeks are not padded; instead they need a rigid structure inside like a rolled-up piece of thin card which springs outwards to keep the knitting well stretched. If you wish, you can also place a disc of card inside the base. The top is closed up by sewing a small disc of dark green crochet inside, above the rolled-up card.

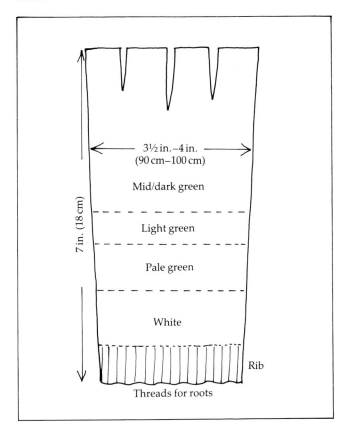

Fig. 158.

Lemon

This is worked in reversed Stocking stitch. Use a lemon-coloured 3-ply yarn and size 2¾ mm needles, and cast on 8 sts. Work 2 rows. On the next row, increase once into every stitch (16 sts). Purl the next row. Then:

1. Next row: increase once into every st. (32 sts).
2. Work three more rows straight, without shaping.
3. Next row: *k1, inc. in next st.,* to last st., k1 (48 sts).
4. Work 15 rows on these 48 sts, ending with a purl row.
5. Next row: *k1, k2tog.,* to end of row (32 sts).
6. Work 5 rows.
7. Next row: k2tog., to end of row (16 sts).
8. Work 3 rows.
9. Next row: k2tog. to end of row (8 sts).
10. Work 5 rows, then thread the 8 sts on to a length of yarn, gather, and sew up, adding Terylene padding as you go.

Button mushrooms (*not illustrated*)

Make these in Stocking stitch and single rib. Tiny mushrooms to add to your collection of vegetables can be made in this way, using beige/fawn/neutral yarn with some brown for the gills in a 2- or 3-ply thickness. With size 2¾ mm needles, cast on 20 sts and work 4 rows in s.s. Next row: K2tog. all along the row to make 10 sts.

Draw these on to a length of yarn and sew up the two side edges. For the gills; case on 20 sts in brown yarn, and work 4 rows in single rib. Do not cast off, but draw the sts up on to a length of yarn and sew up the two side edges.

The stalk; cast on 9 sts and work in s.s. for 6 rows. Sew up the sides to form a tube, attach the drawn-up edge of the gills to the top of the stalk. Now attach the outer edge of the top piece to the outer edge of the gills. Finish off.

Onion

This is simply a Stocking stitch ball, with an extended top and whiskers of yarn at the root end.
(See 'Knitted spheres' on page 110.)

Carrots

Use two 3-ply yarns (double) of two different oranges, on size 3¾ mm needles. They are made in reverse Stocking stitch on 12 sts:
1. Work 22 rows straight.
2. Next row: (k.2tog., k.4,) twice.
3. Next row: purl.
4. Next row: (k.2tog., k.3) twice.
5. Next row: purl.
6. Next row: (k.2tog. k.2) twice.
7. Next row: purl.
8. Last row: (k.2tog.) 3 times (3 sts.) Cast off.
9. Gather up the wide end and sew it together to form the top. Begin sewing the two long edges together to form the point, padding the shape gently as you go. Make a long chain of green yarn and loop this up on the top for foliage, and sew it on firmly.

Fig. 159. A sample showing the crochet ruffle shapes of the kind used on the 'Lichen Log'. Page 46 describes how these are made.

FREE-STANDING STRUCTURES

As knitted and crocheted fabric is soft and pliable, it needs some kind of rigid structure to hold it in place and give it an extra dimension. In garment-making this third dimension is the human body. The garment then becomes something to be seen from all sides, to be moved and changed as the body moves inside it, and as it encounters different lighting conditions. The same appeal is involved in making free-standing pieces, the appeal of being able to see them from all sides in a variety of settings. Anyone who becomes involved in decorative knitting and crochet tends to acquire magpie-like tendencies in their search for all kinds of structures to use as supports for exciting fabrics – cardboard tubes, wire frames and boxes, rings, domes and many weird bits and pieces are mentally 'clothed' in the vivid yarns and textures of knitting and crochet.

Lichen log

The lichen log shown here began life as part of a heavy roll used for transporting newspaper. It was sawn off to the right length, then covered with canvas embroidery, and a similar disc of canvaswork was added to the top and bottom to resemble a sawn-off log. All the joins are hidden by the stitches and by the crochet on top.

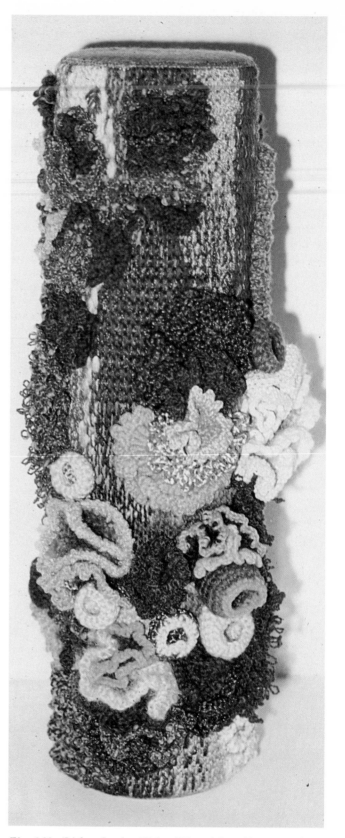

Fig. 160. 'Lichen Log', a 22 in. (56 cm.) log of heavy card covered with canvas embroidery on to which are sewn pieces of crochet lichen, and some hairpin crochet.

Fig. 161. 'The landscape box'.

Landscape box

Some three-dimensional ideas demand that a special container is needed, and often this is more of a problem than making the fabric pieces which cover it. The 'engineering' of three-dimensional structures should be thought out at the same time as the method of fabric-construction; the two are dependent upon each other, and the success of the complete idea depends entirely on the well-thought-out framework upon which the fabric is fixed.

The three-sided landscape box (Fig. 161) is about 10 in. (25.5 cm.) wide across the back and front, and is made of ply-wood strengthened inside with metal corner-brackets. The terylene padding underneath the simple knitted pieces rests on a layer of crumpled chicken-wire which was stapled to the wooden box. The outside of the box is completely covered with green felt which folds well down on to the inside so that the pieces of knitting could easily be sewn to it. The felt was glued into place. The separate pieces of knitting were of the kind which would be produced by young complete beginners – holey, distorted and very rough! These were sewn down on to the hilly structure of wire and padding, and threads of yarn were couched down to represent the river. Pompons for

bushes, and tiny knitted sheep complete the scene.

Curved needles are useful for sewing three-dimensional shapes like these, while cake-boards make useful bases as they accept staples easily. A staple-gun is invaluable!

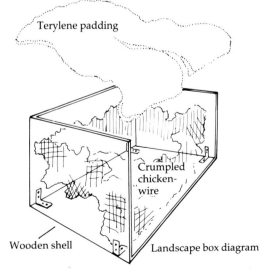

Fig. 162. The box showing the base, wire and padding.

KNITTED SPHERES AND OTHER IDEAS

There are several ways of knitting spherical shapes. The way you choose will depend on the purpose and proposed size, for instance, a small marble-sized ball can be made simply in the one-piece method using a 3- or 4-ply yarn. Thicker yarn and needles will produce one the size of a golf ball using the same pattern. For a larger ball, one could use the 4- or 5-segment pattern which allows you to change the colours more frequently (one for each segment, perhaps?) although one can make stripes quite easily too.

Oval shapes (for eggs, Humpty-Dumptys, rugby balls, lemons, etc.) are produced by extending the central part of the pattern which usually gives several rows to be knitted straight without shaping. When all is said and done, one can even *more* simply knit a rectangle and sew it into a tube, and then gather the top and bottom edges to make a good round shape with only a few creases. This would be perfectly adequate for making tomatoes, apples and doughnuts! Needless to say, if only half the pattern is knitted, you will get a cup or dome shape. A Terylene wadding is most suitable for padding these small shapes to ensure that they will survive rigorous wash-and-dry treatment.

One-piece ball (worked in Stocking stitch)

1. Cast on 6 sts.
2. Row 1: Increase once in every stitch to make 12 sts.
3. 2nd and every alternate row: purl.
4. Row 3: (k1, inc. in next st.) 6 times (18 sts).
5. Row 5: (k2, inc. in next st.) 6 times (24 sts).
6. Row 7: (k3, inc. in next st.) 6 times (30 sts).
7. Row 9: (k3, k2tog.) 6 times (24 sts).
8. Row 11: (k2, k2tog.) 6 times (18 sts).
9. Row 13: (k1, k2tog.) 6 times (12 sts).
10. Row 15: (k2tog.) 6 times (6 sts).
11. Row 16: purl.
12. Draw up the last remaining stitches on to a length of the yarn, and sew up the shape, padding it as you go.

Four-segment ball (worked in Stocking stitch)

1. Cast on 2 sts and knit one row.
2. Increase one st. at each end on alternate rows until there are 8 sts, then work 3 rows without shaping.
3. Increase 1 st. at each end of the next row (10 sts). Work 7 rows without shaping.
4. Decrease 1 st. at each end of the next row (8 sts). Work 3 rows without shaping.
5. Decrease 1 st at each end of every knit row until 2 sts rem. Cast off.
6. Make a total of four segments in this way and sew them together.

Four-segment oval (Egg shape)

This is made in Stocking stitch in the same pattern as the four-segment ball; but work 11 rows without shaping in the centre of each piece, instead of 7 rows.

Five-segment ball

Cast on 2 sts and knit in s.s. as follows:
1. Inc. at each end of the first 2 rows, then on the following 2 *alternate* rows, then on the following 3rd row twice (14 sts). Work 3 rows straight.
2. Next row: inc. 1 st. at each end (16 sts). Work 5 rows straight.
3. Next row: inc. 1 st. at each end (18 sts). Work 16 rows straight.
4. Next row: dec. 1 st. at each end (16 sts). Work 5 rows straight.
5. Next row: dec. 1 st. at each end (14 sts). Work 3 rows straight. Dec 1 st. at each end of next row, following 3rd row twice, following 2 alternate rows twice, then on the next row (2 sts).
6. Next row: purl. K2tog, and fasten off.
7. Work four more sections the same.

Three-dimensional projects may include the following:
Eggs in a basket or egg-box;
Eggs in crocheted egg-cups, with cosies on top;
Coloured balls in a Fair Isle bag as a toy, or marbles 'Gob-Stopper' sweets in a bag;
A bowl of oranges and lemons;
Baked jacket potato on a plate, with melted butter on top;
A large pine-cone;
A chocolate Easter-egg tied with ribbon;
A box of chocolates with lacy sweet cases;
A tray of cacti in pots (see page 49).

Plumose Sea Anemone (see Fig. 165)

The 'Plumose Sea Anemone' is fixed over a structure of shaped chicken wire. The stalks, on to which the plumes are fixed, are knitted and padded tubes. The knitted cover has textured bumps worked into it here and there, supplemented by domes of crochet and large beads. The stalks are attached to a circular ribbed piece of knitting which closes across the top.

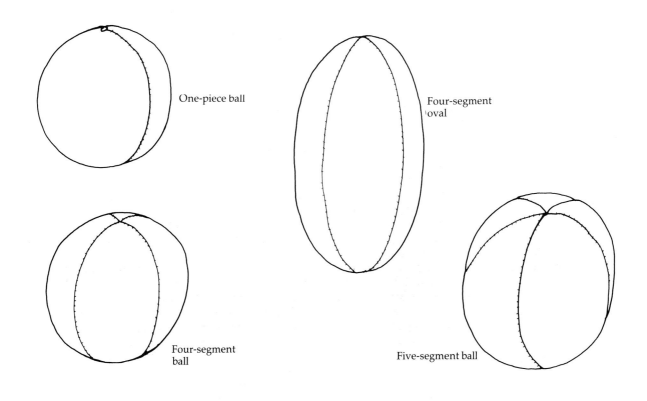

One-piece ball

Four-segment oval

Four-segment ball

Five-segment ball

Fig. 163. Knitted spheres.

Fig. 164. Shapes which can be made from knitted or crocheted spheres.

Fig. 165. 'Plumose Sea Anemone'.

Hanging structures

These hangings are exceptionally easy to make, and very decorative both inside and outside the home. Very ordinary yarns and stitches can be used to create open lacy areas in both crochet and knitting as seen on this one. The hangings can be made to any size: the metal bands upon which the fabric is stretched hold the holes open and allow the light to shine through.

The method of working is partly one of personal preference: one may begin at the bottom and work upwards following a carefully drawn plan or chart, while someone else may like to start somewhere in the middle (which is what I prefer) so that the areas between rings, and the general proportions and balance, dictate what comes next. I work upwards *and*

downwards at the same time, sometimes the top becomes the bottom and sometimes the middle becomes the top!

The rings are incorporated into the work as it progresses, either by sewing them in place or by actually crocheting around them as one works the row where they will be placed. It is not feasible to have the rings loose inside the structure, as the tension of the fabric pushes them askew unless they are fixed. The diameter of the rings is 10 in., 9 in., 6 in. and 5 in. (25.5 cm., 23 cm., 15 cm. and 12.7 cm.) and these can be obtained from craft shops and big stores. Mostly they are made of copper, but wooden ones would do just as well. Heavy beads help to weight the hanging at the

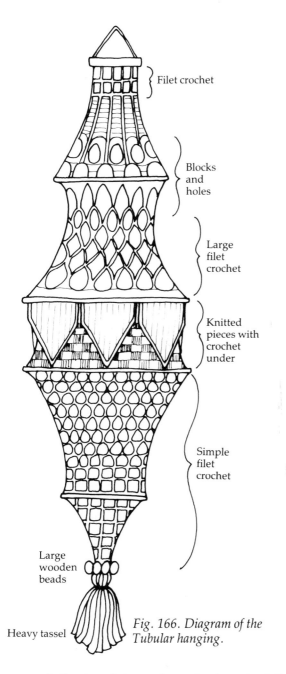

Filet crochet

Blocks
and
holes

Large
filet
crochet

Knitted
pieces with
crochet
under

Simple
filet
crochet

Large
wooden
beads

Heavy tassel

Fig. 166. Diagram of the Tubular hanging.

bottom, and the tassel is an important part of the design as, without it, the effect is disproportionate.

An interesting idea is to have two layers, one inside the other, on one part of the hanging. Here, there are pieces of shaped knitting, made separately, sewn between two rings over an inner layer of crochet. The rings are bound with yarns before they are incorporated into the structure as this helps to fill in any gaps between threads, and also to provide a foundation on to which other pieces may be sewn. Knitted tubular pieces can be made flat, then sewn up one side, or worked on a circular needle, but it is relatively easy to crochet 'in the round', working from one ring to the next with as many stitches as are required. The 'waisted' effect between the rings happens automatically as soon as the structure is held up to take its own weight; the tension which pulls the fabric downwards, while the rings hold it outwards, ensures that it always stays in shape, even after being packed away, or washed and hung out to dry.

Fig. 167. Tubular hanging, overall length 36 in. (92 cm.).

113

Costumes from North America, Panama and Peru. Within one continent, or country, can be found many different styles of costume reflecting the climate and activities of its people. A collection of 'tube figures' could be based on this idea to involve research into the types of costume in, for example, North and South America, using both knitting and crochet.

COSTUME GALLERY (Fig. 168, pages 116–117)

All who have an interest in costume, national or historic, will enjoy making these colourful characters from oddments of yarn, tubes of card, and a little glue. Basically, they are the same as the simple ones on page 41, although these costumes are more detailed, and most of the figures now have arms. The majority of the pieces which make up the costumes are based on rectangles, with simple increase and decrease for shaping here and there. The stitch patterns used are basic, like those used on the other projects, and instructions for making the faces and hair are already described on page 40.

1. The Assyrian king wears a white knitted (s.s.) tunic with several rows of colourful crochet added to the bottom. The waistband is slightly gathered under a Garter st. belt of red and gold. The tunic sleeves are rectangles, folded double and attached at the neckline. They have red cuffs to match the hem. The king's cloak is worn loosely over the shoulders and is made of purple yarn in g.s., with a border of furry yarn to simulate fur, fastening with a large silver button. His gold head-band is a crochet chain of metallic yarn.

2. Benedick is a cavalier who presents a good opportunity to try out tiny pieces of lace-knitting and crochet ruffles. As legs are not possible in this kind of figure, his boots are suggested by a pleat around the tops, and ruffles. The peplum around the base of the bodice is a crocheted strip, and a g.s. sash is tied around the waist to hang down one side. This is decorated with a chain bow. He wears a large white collar of fine white yarn – this could be either loose knitting or crochet. His large brimmed hat is made by crocheting a frill round the edge of a cup-shape, and the feathers are strands of heavy chenille yarn. Alas, Benedick has no arms!

3. The Regency lady wears a high-waisted Empire line dress in two tones of pale blue, slightly wider at the base, with a frill underneath lines of reversed s.s. Her bodice is also knitted in bands of the same stitch, and flat sleeves are attached at the neckline. Her hands are hidden in a tiny muff which is a tube of Garter st. Her bonnet is a crochet dome, quite high, with a brim of flat crochet all round, with bobbles.

4. Miss Prism wears a severe expression and an imposing black bustle of the late 1890s. She has no arms, and all the attention is centred upon the many layers of ruffles of filet crochet in black metallic yarns, sewn into a cascade at the back waist to fall into heavy folds. The skirt piece is gathered towards the lower edges in two places with black ribbon, which ties in bows in the front. A white furry yarn makes an ideal fur collar. Her hair is a ball of grey yarn sewn into a neat bundle and glued to the head, and perched on top of this she wears a black disc of crochet edged with silver.

5. The Green Man is a mythological character, as well-known in the British Isles as he is in many other parts of Europe. He represents the eternal regeneration of the earth, and the seasons of the year. Here he is almost hidden in green foliage; leafshapes cover his green/brown body, green apples (beads) hang from his looped beard and hair, and his crocheted arms have gnarled twig-fingers. Three other similar figures could be made, to represent the seasons of Spring, Autumn and Winter, in the colours most appropriate.

6. The Anglo-Saxon lady is probably the simplest of all these figures, dressed only in a white tunic with an over-skirt of rustic brown. Her wimple is made from two straight pieces of white knitting gathered on top of the head and encircled with a band of gold chain.

7. The Dutch girl is also a very simple figure to make, in a mixture of knitting and crochet. Plenty of bright colours with white sleeves, cap and apron, no shaping is needed on any of these pieces. The sleeves are narrow strips of s.s., and her hands are small pieces of shaped pink card stuck under each cuff.

8. The shepherd wears a traditional smock over his straight tube of brown knitting. The smock, knitted in s.s., has a crochet border (which was added to lengthen it) and small areas of Single Moss stitch to indicate the smocking. He wears a pouch slung across his body, and his hand in his pocket provides a convenient way of attaching his crook to the inside of his sleeve. His arms are folded pieces of s.s. and under his arm he holds a newly-born lamb. The hat is a shallow dome with a circular brim extending from this.

The average height of these costume figures is 11 in. (28 cm).

Key to Costume Gallery
Left to right:
Assyrian king,

Benedick, from Shakespeare's
Much Ado About Nothing

Regency lady

Miss Prism, nineteenth-century England, from
Wilde's *The Importance of being Earnest*

Anglo-Saxon lady   Dutch girl

The Green Man,   Shepherd
a mythological character

LITTLE PEOPLE

The appealing little people seen below on pages 120–1 have guided us into each chapter, and now it is time for them to show us their true colours. They are based on real persons who will certainly recognise themselves and be highly amused! You could do the same, by noting the general stature, clothes, and hair colour of your friends, and presenting the little knitted figures to them on their birthdays. A complete family could be made in this way.

These figures are very small, only about 5½ in. (14 cm.) high, except Alison, who is younger, and all are made on a framework of pipe-cleaners. The clothes have been sewn on to the bodies, so are not removeable, but you can make them in any way you wish. Here are some guidelines to follow.

Basic framework

You will need about 12 pipe-cleaners for each figure: these are used in pairs for extra strength. Twist them together firmly to obtain a shape like that in the diagram.

Pad the body area (indicated by the dotted line) with a thick piece of padding, and hold this in place by binding it around with thick, pink yarn. Use the same yarn to bind tightly round the arms, legs and head, but do not make the limbs too thick as these will be covered by knitted clothes.

Skin covering

When the entire body is covered by yarn, you now need to make the skin-coloured covering for the head and arms. If you wish the figure to wear removeable clothes, it will need an entire 'skin' underneath. This is easily made from small rectangles using fine flesh-coloured yarn on size 2mm–2½mm needles in Stocking stitch. Each piece, instead of being cast off at the finish, has the stitches drawn up on to a length of yarn and gathered into a rounded end which fits on to the ends of the limbs and the top of the head. The gathering thread is then used to sew up the seam. The head and body covering is made from one single piece, gathered round the neck, and with two small holes through which to push the arms. Arm and leg pieces are made separately; the latter may be turned up or left straight as for Alison and Keith.

Skin-coloured yarn

To achieve a fine, smooth skin cover, use the finest yarn and needles you can manage. Fine 3-ply 'baby-yarn' is best, or fine 2-ply Shetland or embroidery yarn. To obtain a good skin-colour, wind off a small skein of cream wool (not synthetics) and tie it up in two places shown in the diagram. This is to prevent it from tangling. Place this in a basin and soak it for about half an hour (or less) in strong tea or coffee, hot or cold, strained or unstrained. Then rinse and dry slowly.

Head covering only

If the shirt or blouse has a high neckline, you need only cover the head and neck. A small, skin-coloured oblong piece with a loose cast-on edge and a gathered top is wrapped around the head and sewn down the back. The seam will be covered by hair or a hat, and features can now be embroidered on to the face.

Arm and leg covering

Again, only simple oblongs are needed for the arms and legs, the length depending on the exact size of your figure and on the length of sleeve, trouser or skirt. It should be measured so that the garment just overlaps the top edge of the 'skin', which should be sewn on to the body framework to keep it secure. If the figure wears shoes or boots, skin-covering will not be needed for the feet. Instead, the colour can be changed at the foot-end of the 'skin' and the shoe-colour knitted on. You will need more stitches for the legs than for the arms, as they are thicker. Large men's boots may be made separately, as the tops need to be pulled over the edge of the trousers but very little shaping is necessary, if any.

Hair

This can be made in a variety of ways, either by making a rectangle of knitting or crochet which is then sewn up to form a cup shape, or by embroidering the hair with straight stitches. Experiment with various yarns; smooth shiny ones are good for straight hair while textured bouclé yarns make ideal curly hair and will disguise any stitches which are used. Sometimes a crochet chain is all that is needed for a line of hair showing underneath a hat. The hair should cover the seam-line down the back of the head. Experiment also with plaits and braids, coiled round the head, and bunches or a pony-tail for a child.

Clothes

Short/bodice pattern. The diagrams show two versions of the top which may be used as shirts for the men, jumpers or dress-bodices. Both will give more or less the same result, but the one based on separate squares is probably easier for beginners to knit. No instructions are given for stitch numbers, as so much depends on the shape and size of the basic figure and the yarn which is used. However, the diagrams are the actual sizes of the figures and clothes shown in the photogrpah, so you may find it useful to aim for the same size (5in.–5½in. or 14cm.). Finishing touches will vary; a little lace collar, a high or polo neck may be crocheted round using a simple d.c. stitch. The opening may be at the back or front, open or closed. The sleeves can be any length, full or narrow.

Trousers. As in most sewing patterns, these are made in two halves with the seams down the centre-back, centre-front and inside legs. The diagram shows how the two rectangles are assembled; make sure you have enough stitches *on each piece* to go round the tops of the legs and half-way round the waist. Use any stitch in knitting or crochet, and measure the pieces against the figure to decide the length.

Skirts. Like the trousers, these need little or no shaping and can be made in any stitch from an oblong. Pauline's short skirt is in double rib to suggest pleats, and Alison's skirt is crocheted in a random-dyed yarn.

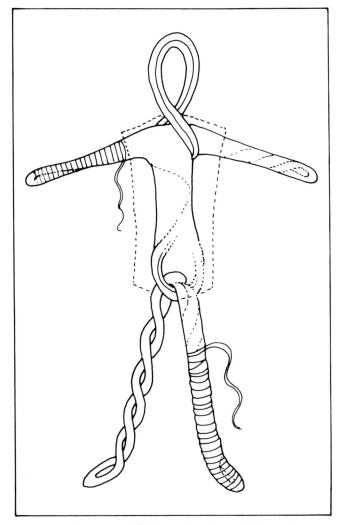

Fig. 169. The Basic Body Shape, showing padding and wrapping.

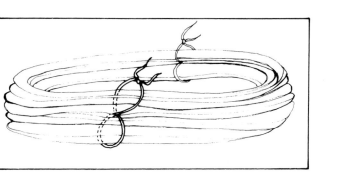

Fig. 170. Tie the skein in a figure of eight before dyeing.

Front

Sleeve

Fold

Neck

Sleeve

Fold

Opening

Back

Version 2

Front

Sleeve

Fold

Neck opening

Sleeve

Fold

Back

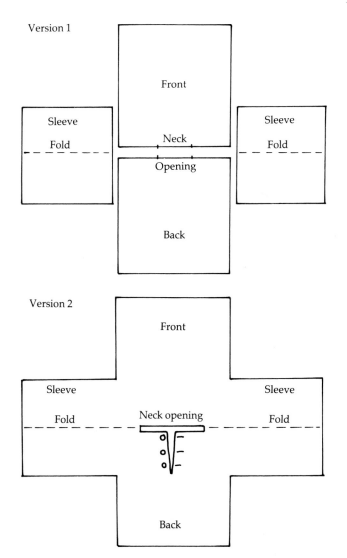
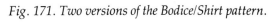

Fig. 171. Two versions of the Bodice/Shirt pattern.

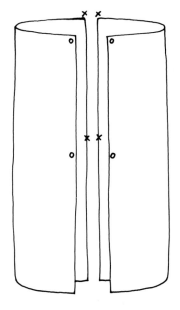

Fig. 172. The Trouser pattern: join x to x and o to o.

119

Little people – key to characters

1. The shepherd is knitting a black woolly jumper for his pet black sheep, as he has done for the three white ones.

2. Robin the Gnome knits his other red boot and hopes he will not be needed before it is finished.

3. Alec enjoys fishing and so is knitting some basic equipment. There's a plaice for us!

4. James is a crochet expert and so is now learning to knit. The trouble is, he has not yet seen the diagram on page 22.

5. Sylvia is really a fairy who makes beautiful lacy garments from cobwebs and dreams and whispers.

6. Keith is making a colourful crochet waistcoat for his wife. He is strategically placed between the two expert ladies.

7. Pauline is crocheting a bag in which she will keep her rainbow-coloured reflections.

8. Alison is ready to begin her knitting, if only the dog would leave her wool alone.

Sylvia's long skirt is also crocheted, with a scalloped edge, and the top is gathered on to the bodice.

Shepherd's smock. The front and back were made all-in-one as in Version 2 of the shirt/bodice pattern, except that the smock is larger. Beginning at the lower edge, more stitches are needed to give the fuller appearance, and these are then decreased towards the top shoulder area. Change to a single rib here to look like gathers until well over the back, when you will need to increase again for the fuller back part. The neck opening is the same as for Version 2 of the shirt, but here a large collar is made separately and added, leaving a small gap at the front. The two sleeves were made, in this instance, by picking up stitches from the shoulder/sleeve edges of the smock and knitting towards the cuffs. They could also be made separately if preferred.

Hats. In each case, the main part is knitted, beginning at the widest edge of the *crown* and, after sewing this part up, the brim is crocheted on. About 26–28 sts were needed, using fine yarn and needles, decreasing at regular intervals along the row, over 8–10 rows. The last 6–8 sts were gathered on to a thread to form the top. The sou'wester (which Alec wears) has a longer crochet brim at the back than at the front and the shepherd's hat has a chained band around the crown. The gnome's hat has an extended point which hangs down and ends with a tassel.

Accessories. The knitting needles and crochet hooks used by the little figures are made from half-cocktail sticks, with wooden bead knobs. The shepherd's crook is made from two pipe-cleaners, tightly wrapped with yarn.

Dog. To make the dog, work along the same lines as the sheep diagram on page 61, but make it slightly smaller. The body-covering can be made in any 'dog colour' from a series of tiny rectangles, the main body piece being gathered at the nose and tail ends. The ears and tail are made separately, and tiny coloured patches may be sewn on afterwards.

Rainbow bags. Rectangles again! Knitted in Stocking stitch, with 3- or 4-ply yarns on size 2¾ mm needles. 30 sts – 4 rows each of seven rainbow colours, sewn up along the base and one side, gathered at the top with a chain draw-string and filled with wool.

LARGE PEOPLE

Historical costume figures are exciting and rewarding to make, as the research involved, and the inventiveness required to achieve an authentic-looking result, are two areas which require attention to detail and an interest in history. While on one hand it is easy to believe that a cursory look at the costume of one period will enable you to make a fair replica, you will soon discover that more than a polite familiarity is needed to make all the various parts of that costume in knitting or crochet, as these bits have to be seen from all sides and angles, and must therefore be convincing.

The figure of Queen Elizabeth I was based on the famous portrait by the miniaturist Nicholas Hilliard, which is known as the 'Phoenix Portrait' because of the Phoenix pendant hanging on her bodice; a symbol of regeneration. As this portrait is not full-length, but shows only a little more than half of the figure, it was necessary to look at many other portraits of the same period to fill in details which were missing. This model

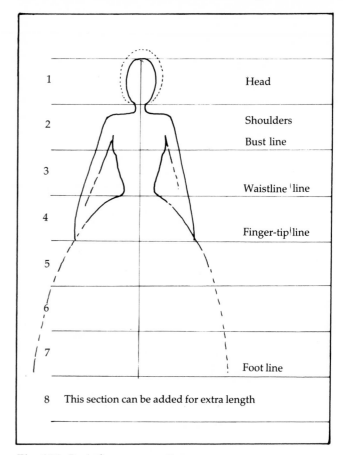

Fig. 174. Basic figure proportions.
The various parts of the body fit into seven partitions. A man may be somewhat taller, so extra length may be added in the eighth partition at the bottom, keeping the other parts in the same place and in the same proportions, except for the arms which can be made a fraction longer. (In tall people it is usually only the arms and legs which are noticeably longer.)

Fig. 175. Queen Elizabeth I, based on a portrait by Nicholas Hilliard.

is a mere 13 in. (32 cm.) tall, and so ways had to be found to keep the essential spirit and 'feel' of the portrait while at the same time keeping the tiny details *in scale*.

Even fine yarn on fine needles produces a texture which is difficult to hide when trying to reproduce smooth skin, satin, or other low-textured fabrics, and tiny shapes are equally problematical for similar reasons. Indeed, *scale* seemed to be the all-important factor; anything too large, too bold or too textured would soon overwhelm the over-all effect of the minutely detailed portrait. It soon became clear that every detail could not be fitted in: on this scale, an impression of richness would have to be enough.

After studying the portrait very closely, one is obliged to make a working plan, to avoid doing things in the wrong order and so having to remove them to put something underneath, etc. . . . The following notes are intended to suggest an order of work; they do not explain how to make this particular costume, *only the basic figure*, for everyone will have their own ideas about which character they would like to make.

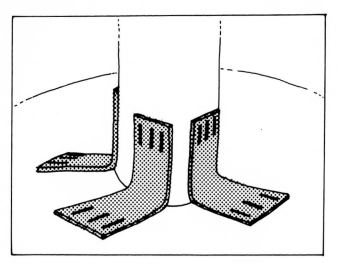

Fig. 176. The base and centre-piece upon which the body is built are kept rigid with angled card, firmly stapled down.

WORKING NOTES

Collect as much information as you can about the character you wish to make, photographs, drawings, etc., and keep these by you all the way through the project. Find as many other pictures as possible of other costumes from the same period, and of the same style, and study these carefully. Make extensive drawings of every part of the costume you can see, using bits from all sources to fill in gaps which are not clear. This drawing stage (however inexpert) is essential to clarify the way in which a costume is constructed. You will need to know what the back looks like, how the collar is fixed on to the bodice, what those vague bumps on the shoulders are. To draw all these things is an aid in making them correctly.

Yarns. Give yourself a break, at this point, and begin to accumulate the yarns needed for your project. You will need good daylight for choosing colours, as artificial light changes them out of recognition. Because of the scale of the model (even if it is bigger than this one) only the finest yarns are best, to keep the stitch textures in scale. If possible, choose 2-ply, 3-ply and 4-ply yarns. Match the colours up as carefully as possible; using the correct kind of white (not the brilliant white of synthetic yarn, but cream or off-white) and choose the slightly softened, faded colours of historical portraits rather than the garish primary colours of the cheaper knitting yarns. Stranded embroidery yarns, cottons, linen, and any other fine threads are all useful, as are metallic threads. Use anything which will produce the effect you seek.

Other useful additions. All kinds of other bits and pieces may be needed to imitate real jewels and embellishments such as buckles, swords, flowers and fans. This is the time to search through the button-box and the 'bit-box' for anything which might be used – tiny buttons, beads, sequins, braid, ribbon, cord, jewellery mounts, soft wire, artificial leaves and bits of fine chain. Craft shops stock a variety of useful things in this line. Even cocktail and tiny lollipop sticks, cardboard tubes and metal rings, washers and bits of coloured acetate can be used as a basis for 'props'. But remember that it is the scale of these accessories that is important.

Body base

Historic costume figures wearing full-length gowns do not need legs, and so the only parts of the basic body shape which usually concern us are those above the waistline. A useful base-board to use is a cake-board, or a thick cork mat like the kind used by flower arrangers. The central core of the body is a thick piece of broom handle set into a hole in the base firmly glued and strengthened by card supports (see the diagram) and stapled in place.

For the arms and shoulders, firm wire is doubled *twice* and twisted, then wrapped around the broom handle at the shoulders below the neck. Make the arms long enough to reach the required proportional length, to the finger-tip line. This wire is held in place when padding and wrapping begins.

Body padding

Use strips of thick terylene padding to pad the head and body by winding it round and round to achieve the required thickness. Use this to hold the wire arms in place, wrap down the arms making them thinner at the bottom. Do not overpad the neck area. Hold the padding in place, and mould it into shape by wrapping with thick pink yarn, or strips of soft pink fabric, while at the same time shaping the head, shoulders and body. It is not usually necessary to pad further down than the waistline, but this will depend on the type of figure you are making.

Skin covering

This is only needed in places where the skin actually shows, e.g. face and hands, so small finely-knitted oblongs can be used for this. For the head, the piece of knitting should be long enough to reach from the top of the head to well below the neck. It should stretch well round the head but need not meet at the back as the hair and collar will usually cover this part. Gather it into the neck by running several threads around it and easing the fullness.

The face may be gently moulded when the head covering is in place. Pinch the bridge of the nose hard to make a dent into the corner of the eye-socket and carefully stitch behind this area, from one eye-corner to the other, making the bridge of the nose slightly prominent. Use the matching yarn for this, and tiny stitches which show as little as possible. Nip the nose into a ridge all the way down, and gently sew underneath this from one side to the other to keep it in place. To set the eyes further into the head, make a knot in the yarn at the end; then, from the *back* of the head, stab straight through the head towards the eye, bringing the needle out to one side of the bridge of the nose. Enter the needle again at this point and stab straight back towards the knot, pulling firmly to indent the skin-covering. Do this as many times as necessary to make it look realistic, and repeat it for the other eye.

Hair

Ideas for hairstyles have already been discussed in connection with the other knitted and crocheted figures, and this project is only a larger version of those. However, there is probably more scope here for a more interesting use of the various stitches, such as ribbing (which will produce ridges) and Moss stitch (which will look like tiny curls), generally producing a more precise style than with one of the smaller characters. Crocheted cups and domes are very useful for making caps of hair, to which can be added ringlets and other details around the edges.

Glossary

Only the terms and abbreviations which have been used in this book are included in the following list:

KNITTING ABBREVIATIONS

alt.	alternately
beg.	beginning
cm.	centimetres
dec.	decrease
d.k.	double knitting yarn
g.st.	Garter stitch
in.	inches
inc.	increase
k.	knit
p.	purl
patt.	pattern
p.s.s.o.	pass the slipped stitch over
p.wise	purl-wise, i.e. with the yarn in front
rem.	remaining
rep.	repeat
R.S.	right side
sl.st.	slip stitch, i.e. without knitting it
st.	stitch
s.s.	Stocking stitch
tog.	together
w.r.	wrong side
y. fwd.	yarn forward
y.r.n.	yarn round needle

Reversed Stocking stitch is the other side of ordinary Stocking stitch, sometimes called the 'purl side'. Ridges of reversed s.s. are made by purling on a knit row, and by knitting on a purl row.

CROCHET ABBREVIATIONS

beg.	beginning
ch.	chain
ch.sp.	chain space
d.c.	double crochet
d.tr.	double treble
dec.	decrease
h.tr.	half treble
inc.	increase
rep.	repeat
s.s.	slip stitch (see below)
st.	stitch
tog.	together
tr.	treble
y.o.	yarn over

A 'foundation chain' is the chain on to which the first row, or round, of crochet stitches is made.

A 'slip stitch' is a means of connecting two stitches or sections together without making a new stitch in the process. It is done by entering the hook into a chain space, or another stitch, and yarning over, then pulling the y.o. straight through – including the loop on the hook.

GENERAL

Brackets () means that whatever instructions are written inside them should be repeated by however many times it says outside them: e.g. (k.2tog., k3.) 4 times.

* The asterisk is similar. It encloses a set of instructions which have to be repeated, not necessarily on one row only, but sometimes over several rows.

A motif is a single unit of design. A border pattern repeats a motif in a long line, and an all-over pattern repeats a motif over a large area. A motif may either be used alone, or with others, or repeats of itself.

Index

OTHER CRAFT TITLES

THE MINIATURE WORLD OF PRESSED FLOWERS
by Nona Pettersen

The miniature scale is ideal for the aspiring creating flower arranger, and a stimulating one for those familiar with the usual range of flowers and composition. This attractive, enthusiastic book also explores the plentiful source of plant life, suitable for miniature work, that is surprisingly available in urban areas.

QUICK & EASY WOODEN TOYS
by Alan Pinder

How simple wooden toys can be made easily, quickly and cheaply, without the aid of an expensive workshop and using only wooden fittings. A practical and enjoyable book which combines clear instructional text, colour pictures, diagrams, and exploded drawings.

EVERY KIND OF PATCHWORK
edited by Kit Pyman

'Really lives up to its title, and is sufficiently easy to follow that even the most helpless needleperson would be tempted to have a go. But there's plenty, too, for the experienced.'
The Guardian.
Cased and Paperback.

EVERY KIND OF SMOCKING
edited by Kit Pyman

The description of the basic technique is followed by sections on children's clothes, fashion smocking, experimental smocking and creative ideas for finishing touches.

THE CHRISTMAS CRAFTS BOOK

Creative ideas and designs for the whole family to make objects with a Christmas flavour: table and room decorations, stars, Christmas tree ornaments, candles and candlesticks, angels, nativity scenes, paper chains and Christmas cards.

MADE TO TREASURE
Embroideries for all Occasions
edited by Kit Pyman.

This book offers a rich variety of ideas for embroideries to be made to commemorate special occasions – christenings, weddings, birthdays – from simple greetings cards to a gold-work panel for a golden wedding. A heart warming present precisely because it is specially made.

THE SPLENDID SOFT TOY BOOK

The Splendid Soft Toy Book contains a wealth of ideas and pictures for making a wide variety of toys and dolls, from a green corduroy crocodile to detailed traditional, even collectors' dolls. More than 60 full colour pictures and over 70 black and white illustrations show the reader how to fashion appealing figures and animals of all shapes and sizes.
Cased and paperback.

THE ART OF PAINTING ON SILK
VOLUME 1
edited by Pam Dawson

This book describes the basic techniques, tools and materials, required to paint on silk, followed by colourful examples of designs and finished items that will be useful to both the beginner and experienced artist. Useful motifs to trace are given in the final section of this book.

THE ART OF PAINTING ON SILK
VOLUME 2 Soft Furnishings
edited by Pam Dawson

A colourful and inspirational book which covers a wide range of soft furnishing designs, from cushions and wall hangings to bed covers and lampshades. Each design is shown in full colour together with details of materials required, methods used and simple to follow charts of the painted motifs. Whether you are an experienced artist or a complete novice it is possible to learn from this book.

If you are interested in any of the above books or any of the art and craft titles published by Search Press please send for free catalogue to: Search Press Ltd., Dept B, Wellwood, North Farm Road, Tunbridge Wells, Kent. TN2 3DR.